Digging Into Autodesk Land Desktop 2004
Level 1 Training

Rick Ellis

Software Solutions, Inc.

PO Box 344
Canby Oregon 97013
www.cadapult-software.com
training@cadapult-software.com
(503) 829-8929

ISBN 0-9740814-1-8

Cadapult Software Solutions, Inc.
PO Box 344
Canby Oregon 97013
www.cadapult-software.com
training@cadapult-software.com
(503) 829-8929

About the Author

Rick Ellis is the founder and CEO of Cadapult Software Solutions, Inc. At Cadapult Software Solutions, Rick provides training and consulting services to clients around the country helping them to get the most out of their design software investment. He is also an independent consultant and instructor specializing in Autodesk Land Desktop, Autodesk Civil Design, Autodesk Survey, Autodesk Map, Autodesk Raster Design, AutoCAD, and Bentley MX Civil Design software. Mr. Ellis is a member of the Autodesk Developer Network, co-author of "Introducing Autodesk Civil 3D", a K-TEK Certified Professional (KCP), a select author for CAD Digest, an instructor for the AUGI training program, as well as a Level I and Level II Bentley MX certified instructor.

Prior to founding Cadapult Software Solutions Rick worked as a CAD Manager and Civil Designer and then as the Technical Services Manager for an Autodesk Reseller. He brings this real world experience and industry knowledge to his training and consulting projects to provide practical examples and solutions to clients.

You can email Rick at: rick@cadapult-software.com

About Cadapult Software Solutions, Inc.
www.cadapult-software.com

Founded in 2002, Cadapult Software Solutions, Inc. is an independently owned small business located near Portland, Oregon specializing in training, consulting services, and technical support for CAD systems with a focus on the Civil/Survey/GIS industry. Cadapult Software Solutions helps clients maximize the return on their software investment through training classes, consulting services, and support. We offer a wide range of Training options, from standard open enrollment classes to customized on-site training. Our mobile training lab gives us the flexibility to bring classes to our clients regardless of the location. Support options ranging from telephone support to on-site visits help to ensure the continued success of your CAD solution. Although we hold several certifications with both Autodesk and Bentley, Cadapult Software Solutions is an independent company and therefore can provide recommendations and solutions that best fit a clients needs rather than being limited to a specific company's product line. Further affiliations with other consultants and software companies give Cadapult Software Solutions a broad range of experience and industry knowledge to draw from that is not common for a company of it's size.

Acknowledgements

I would like to thank all the people who made this book possible. There is no way that I can adequately explain the importance of their involvement. But I can simply state that this project could not have been completed without any of them.

Many thanks to Russell Martin for his countless hours of help during the technical editing process. His input and comments were invaluable in filling in the gaps and turning this book into a complete finished product.

Thank you to Brandt Melick and the City of Springfield, Oregon for providing the project data and their input into the process and topics covered.

Thank you to Ken Baker and B&B Print Source for their help and patience printing the book.

And most of all thank you to my wife Katie and our children Courteney, Lucas, and Thomas. Without their love, support, and understanding this book would never have been possible.

Introduction

Welcome to **Digging Into Land Desktop 2004**. This book of tutorials is designed to introduce you to the fundamental concepts and procedures commonly used in Autodesk Land Desktop. It is meant to be a resource and a supplement to instructor led training.

It is not meant to, nor is any book able to, replace instructor led training. It is also not a book that will teach you civil engineering, although you may pick up some concepts along the way. This book is the material that I use for Level 1 Land Desktop classes that I teach. It works through a basic Land Desktop project from beginning to end. Showing you many different methods of using Land Desktop to accomplish certain tasks and solve problems along the way. Keep in mind that all projects are different so this is not the exact process that you will use to go through every project. It is more important that you understand the individual tools that we are using than focusing on the overall process. This way when you encounter examples that we do not cover in this book you can identify and use the proper tool to accomplish your task.

Included Data

The data supplied on the CD with this book as well as the exercises are designed to work with a standard "out of the box" installation of Land Desktop 2004. The dataset for chapter one, which when extracted to your C drive will create a folder called "LDT level 1" in a folder called "Cadapult Training Data", contains all of the data needed to complete all of the exercises in the book. The drawing files contained in this first dataset are an AutoCAD 2000 format so that you could complete the exercises with any version of Land Desktop that is at least version 2 or higher. The remaining datasets are all in the AutoCAD 2004 format and are snapshots of where your project should be at the beginning of each chapter. This will allow you to jump in at the beginning of any chapter of the book and do just the specific exercises that you want to do if you do not have time to work through the book from cover to cover.

Style Conventions

Pull-down menus: **Map** ⇒ **Utilities** ⇒ **Project Workspace**
Command line entries: `Command: Zoom`
Buttons: **<<New>>**

Table of Contents

Chapter 1

Project Setup

In this chapter we will describe Autodesk Land Desktop (LDT), Autodesk Civil Design, and Autodesk Survey. We will also introduce how Land Desktop works with external project databases as well as the two different methods of launching Land Desktop and when to use each one. Finally we will setup a new Land Desktop project and create our first Land Desktop drawing, which we will use throughout the book.

- **Description of Autodesk Land Desktop, Autodesk Civil Design and Autodesk Survey**

- **Description of Project Data**

- **Deciding how to launch the Program**

- **User Interface**

- **Opening and Setting up a New Project**

Dataset:

To start this chapter we need to extract the dataset named **"Chapter 1 Cadapult Level 1 Training.zip"** from the CD that came with the book. If you extract it to your C drive it will create a folder named **C:\Cadapult Training Data\LDT Level 1** which will contain all of the base data that we will use for the exercises in this book.

1.1   Description of Autodesk Land Desktop, Autodesk Civil Design and Autodesk Survey

Land Desktop is Autodesk's version of AutoCAD for Civil Engineers, Surveyors, and anyone who works with them. This means that it is not a program that works with AutoCAD but instead AutoCAD is part of Land Desktop. Actually, Land Desktop includes AutoCAD, Autodesk Map as well as the ability to work with Points, Alignments, COGO, Labeling, Parcels, Surfaces and Quantities. You may have Autodesk Survey and/or Autodesk Civil Design in addition to Land Desktop. Autodesk Survey gives you the ability to communicate with your survey data collector, adjust a traverse and import linework generated in the field. Autodesk Civil Design adds the ability to work with Profiles, Cross Section, Hydrology, Pipeworks, Grading and Sheet Manager.

1.2   Description of Project Data (External Databases)

When working in LDT you are not just working with an AutoCAD drawing you are also working with an external project. This project is composed of several databases that store different types of your civil project data such as Points and Surfaces. Each project can have several drawings attached to it but a drawing may only be attached to one project. The project data is what Land Desktop uses for all of its calculations such as determining quantities and creating profiles and cross sections. So at times the data in the project files is more important than the data in your drawings. If you create a good clean set of project data then the drawings can easily be generated from it. For this reason as you are working in Land Desktop it is important to consider not just what your commands are doing to the graphics on the screen but also what they are doing to the project data. One thing that you must insure is that the project and the drawing stay in synchronization. Having different data in the project than is reflected in the drawing is where many common problems begin.

1.3   Deciding how to launch the Program (Land Enabled Map v. Land Desktop)

When using Land Desktop you are required to have a project attached to your drawing. If you want to work in AutoCAD without having a project attached you can do so by launching a version of Autodesk Map from the Icon titled Autodesk Land Enabled Map. This will launch Autodesk Map without requiring you to attach a project. However, you will not have the use of any of the Civil/Survey tools. The other advantage of Land Enabled Autodesk Map is that you will be working in MDE or Multiple Design Environment, which allows you to open more than one drawing at a time.

MDE is not available in Land Desktop. An example of when Autodesk Map might be the better choice is if you are working on standard detail, where you would benefit from MDE, but wouldn't need the use of Civil/Survey tools.

So you will have two icons to launch the program from, 1) Land Desktop for Civil/Survey work that will utilize your project data and 2) Land Enabled Map for regular AutoCAD work without the project data, see examples below.

Autodesk Land
Desktop 2004

Autodesk Land
Enabled Map

1.4   User Interface

The user interface of Land Desktop is very similar to AutoCAD, with a few additions. You can reposition and customize tool bars and pull-down menus the same way you would in AutoCAD. However, there are a few additional features that you will need to become acquainted with.

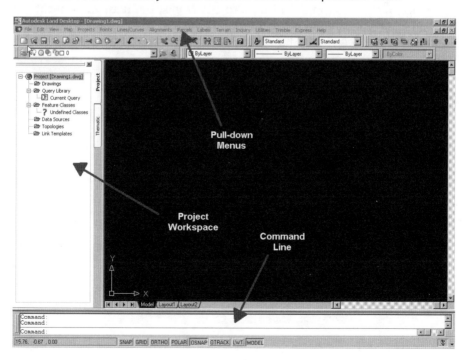

1.4.1 The Project Workspace

The Project Workspace, or as many people refer to it "the big white box on the left side of my screen in Land Desktop or Autodesk Map", is an area that displays information about your Map project and allows you to execute many Map commands with a right-click in the appropriate area. It is a very useful and productive tool if you are working with the Map commands such as attaching drawings or external databases and working with queries or topologies. However, if you are not using those commands then the Project Workspace consumes a large amount of screen area. This has been improved in Map 2004 with the addition of transparent and hiding options enabled with the Project Workspace.

Once you decide that you do not need the Project Workspace displayed, either short or long term, you can turn it off. There are two ways to remove the Project Workspace from the screen. You can simply close the window by clicking the "x" in the upper right corner of the box or by right-clicking in the Project Workspace and selecting Hide. This will immediately close the window. However, it will reappear the next time that you launch the program. This is because the Project Workspace is not controlled by the AutoCAD profile like toolbars and pull-down menus, instead it is controlled by the Map options which are written to the acadmap.ini file. So to turn off the Project Workspace until you want it back on again you need to go to the Map Options. Select Map $\Rightarrow$ Options. Then clear the check box "Show Workspace on startup" on the Workspace tab.

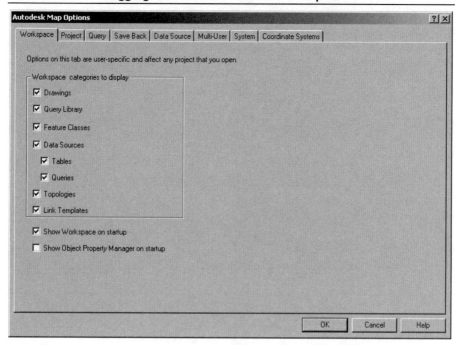

This will turn off the Project Workspace the next time you launch the program. If you don't want to restart you can just close the Project Workspace at this time.

For the exercises coming up in chapter two we will be using the Project Workspace so if you have turned it off you should restore it by selecting Map ⇒ Utilities ⇒ Project Workspace.

1.4.2   Menu Palettes

When Land Desktop is combined with Autodesk Survey and Autodesk Civil
Design there are more pull-down menus available than will fit across the
top of most people's screens. So Land Desktop uses a system of Menu
Palettes to manage the pull-down menus. In this book we will work with
the standard menu palettes that come with the program. However, you
can easily customize them by arranging the pull-down menus the way you
want them with the MENULOAD command and then saving that
configuration in the Menu Palette Manager.

To change between menu palettes select **Projects** ⇒ **Menu Palettes**.

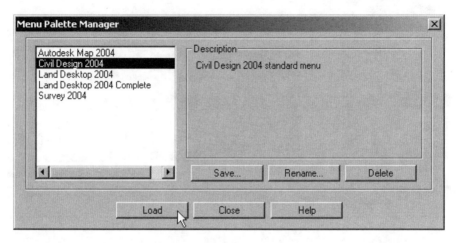

In the Menu Palette Manager select **Civil Design 2004** and then click
<<Load>>.

This will load the Civil Design pull-down menus as well as some other
selected pull-down menus from AutoCAD, Map, and Land Desktop. After
reviewing the changes switch back to the **Land Desktop 2004** menu
palette.

1.5   Opening and Setting up a New Project

When you launch Land Desktop you will see that the **Start Up** dialog box has options for Open, New and the Project Manager. The Open and New options in LDT are both Project Based.

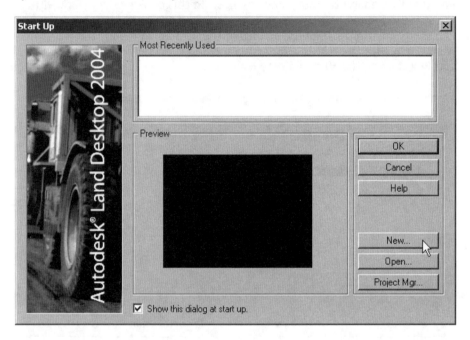

1.   Click the **<<New>>** button.

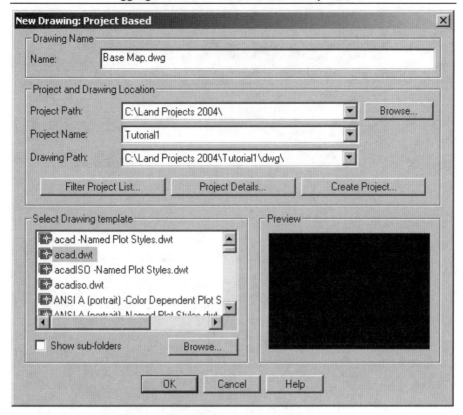

2. Name the drawing **"Base Map.dwg"**.

You do not need to type ".dwg" it will be added for you.

3. Set the Project Path: **C:\Land Projects 2004**.

This path should not include the project name. It is a root directory that will contain many projects. This path should not be just a drive letter such as C:\, it needs to contain at least one subdirectory. Although the software will allow you to use the root of the drive as the project path, there are some known issues, so this practice should be avoided.

4. Then click the **<<Create Project>>** button.

5. Set the Prototype: **"Default (Feet)"**.

This prototype contains settings and defaults for our entire project. These settings can be modified later during the project as needed.

6. Set project Name: **"Cadapult Level 1 Training"**.

7. Leave **Drawing Path for this Project** set to "Project DWG Folder".

This is the default location for the drawing files associated with this project. The drawings are not required to be in this folder. Also, all drawings in this folder are not automatically associated with this project.

8. Click the **<<OK>>** button.

This will bring you back to the New Drawing dialog box where you should select a drawing template.

9. Select the template **"acad.dwt"**.

The New Drawing dialog box should look like the figure below when it is completed.

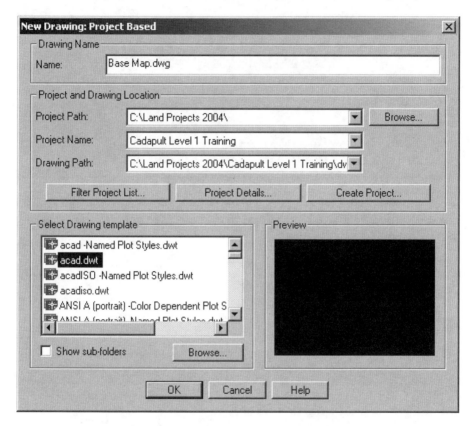

10. Click **<<OK>>** to create the drawing.

11. Answer **<<No>>** if asked to save changes to Drawing 1.

12. When the **Create Point Database** dialog box appears, click **<<OK>>** to create the Point Database using the default values.

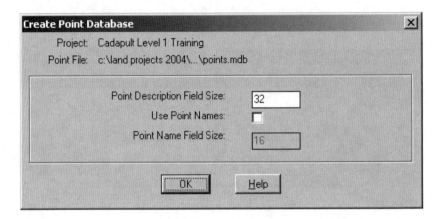

If you will have point descriptions longer than 32 characters you can adjust the description field size here. Point Names are used only if you will be using alphanumeric point numbers.

13. Next the **Drawing Setup Wizard** will begin.

These settings must be set for each drawing in your project.

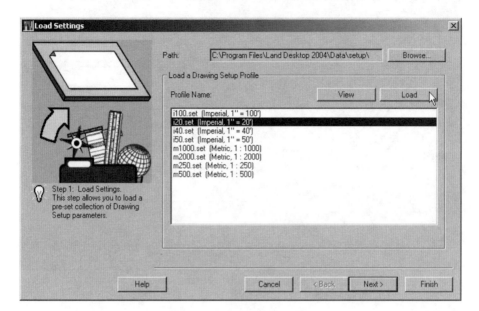

14. Select the profile **"i20.set (Imperial, 1" = 20')"** and click **<<Load>>**.

15. Click **<<Next>>** to move to the **Units** panel.

Here you specify the units and display precision for this drawing.

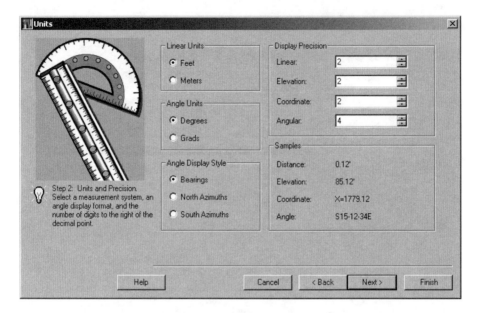

16. No changes are needed.

17. Click **<<Next>>** to move the **Scale** panel.

Here you specify the horizontal and vertical scale of the drawing. The horizontal scale should be the scale you plan on using to plot this drawing. The horizontal scale also controls the Dimension Scale and Linetype Scale of your drawing. The vertical sale controls the vertical exaggeration of your profiles and cross sections.

> 18. Confirm that the Horizontal Scale is set to **1" = 20'** and set the Vertical scale to **1" = 4'**.

This will give us a 5:1 exaggeration for our profiles and cross sections.

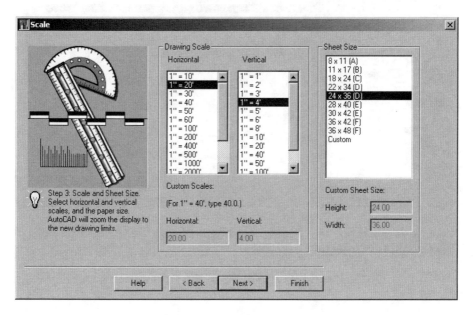

> 19. Click **<<Next>>** to move to the **Zone** panel.

20. Select the Category **"USA, Oregon"**.

21. Select the Zone **"NAD83 Oregon State Planes (Polyconic), South Zone, Intn'l Foot"**.

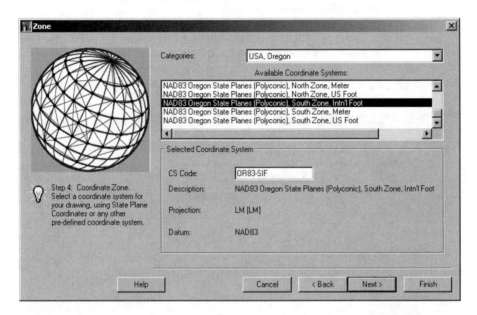

If your drawing is not on a known coordinate system use the Category of "No Datum, No Projection".

22. Click **<<Next>>** to move to the **Orientation** panel.

Use this if you want to apply a north rotation to your drawing. If you do this the x,y coordinates will be different than your Northing and Easting coordinates.

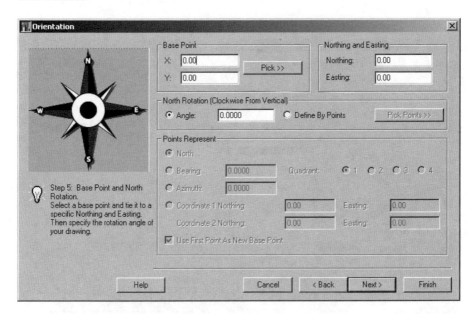

23. Make no changes to the Orientation Panel of the wizard for our project.

24. Click **<<Next>>** to move to the **Text Style** panel.

Here you will load text styles for your drawing. These are AutoCAD text styles with heights. Those heights combined with the horizontal drawing scale you selected earlier will determine the height of the text in the drawing.

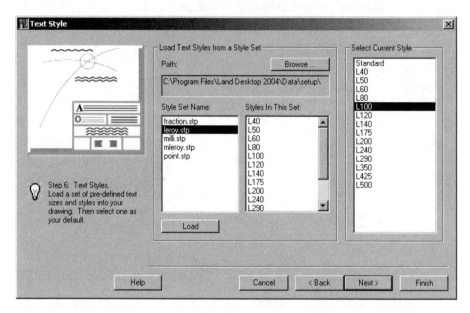

25. Confirm that **L100** is set as the **Current Text Style**.

26. Click **<<Next>>** to move to the **Border** panel.

These settings will automatically draw or insert a border in to model space if you wish to use them.

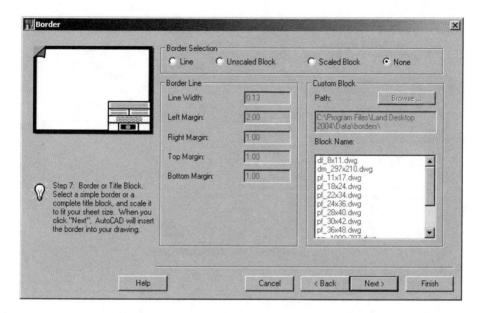

27. Make no changes to the Border Panel of the wizard for our project.

28. Click **<<Next>>** to move to the **Save Settings** panel.

If you plan to use these same settings as a standard on future drawings type in a Profile Name and click **Save**.

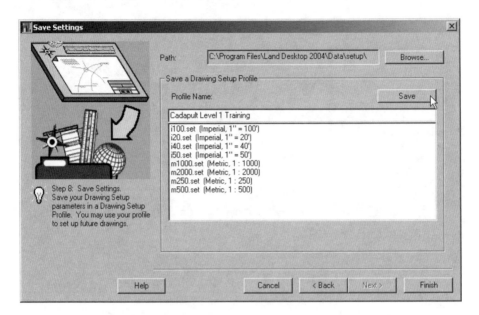

29. Enter a Profile Name of **Cadapult Level 1 Training** and click the **<<Save>>** button.

30. Now click **<<Finish>>** to complete the wizard.

You will then be shown a dialog box that summarizes what you have set up in the wizard.

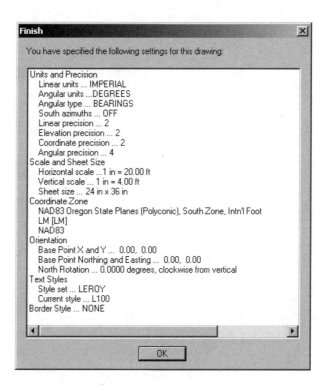

Finish

You have specified the following settings for this drawing:

Units and Precision
    Linear units ... IMPERIAL
    Angular units ...DEGREES
    Angular type ... BEARINGS
    South azimuths ... OFF
    Linear precision ... 2
    Elevation precision ... 2
    Coordinate precision ... 2
    Angular precision ... 4
Scale and Sheet Size
    Horizontal scale ...1 in = 20.00 ft
    Vertical scale ... 1 in = 4.00 ft
    Sheet size ... 24 in x 36 in
Coordinate Zone
    NAD83 Oregon State Planes (Polyconic), South Zone, Intn'l Foot
    LM [LM]
    NAD83
Orientation
    Base Point X and Y ... 0.00, 0.00
    Base Point Northing and Easting ... 0.00, 0.00
    North Rotation ... 0.0000 degrees, clockwise from vertical
Text Styles
    Style set ... LEROY
    Current style ... L100
Border Style ... NONE

OK

31. After reviewing your new settings select <<OK>>.

You have now created a new Land Desktop Drawing that is attached to our Project. If you need to make any changes to the setup information that you set in the Drawing Setup Wizard select **Projects ⇒ Drawing Setup.**

1.6   Chapter Summary

In this chapter we have gotten acquainted with Land Desktop, its interface, and the concept of the external project database. We also have setup a new project and a new drawing. As we proceed through the following chapters you will see how important this initial setup is to working effectively in Land Desktop. You will find that if you understand Land Desktop's settings, and learn to use them properly, that all of the steps and processes that you need to perform for design and drafting are much easier.

Chapter 2

Data Collection and Base Map Preparation

This section shows how to build base maps with existing map information. This section covers importing GIS data, inserting registered ortho photography, as well as attaching and querying specific layers from planimetric map sheets and Digital Terrain Model (DTM) source sheets.

- **Importing GIS Data**

- **Using Queries to Manage and Share Data**

- **Inserting Images And The Project Boundary**

- **Importing Point And Breakline Information From Aerial Mapping**

Dataset:

To start this chapter we will continue working in the drawing named **Base Map.dwg**. You can continue with the drawing and project data that you currently have from the end of the previous chapter or you can install the dataset named **"Chapter 2 Cadapult Level 1 Training.zip"** from the CD that came with the book. Extract the dataset to the folder **C:\Land Projects 2004** or whatever folder you have been using as your project path. Extracting the project from the dataset provided will ensure that you have the project and drawings set up correctly for the exercises in the following chapter and overwrite any mistakes that you may have made in previous exercises.

2.1 Importing GIS Data

In the previous exercises we created a project and a drawing. In this chapter we will add parcel and street data from the GIS department as well as an aerial photo of the project area. We will also add Point data and breakline data that we will use later to build a surface.

2.1.1 Importing ESRI Shape Files

1. Open the drawing **Base Map.dwg** from the project **Cadapult Level 1 Training** if it is not already open.

2. From the main menu select **Map** ⇒ **Tools** ⇒ **Import**.

3. Browse to:
    C:\Cadapult Training Data\LDT Level 1\GIS Data

4. Set the Files of type to **ESRI Shape**.

5. Click **<<OK>>** to continue.

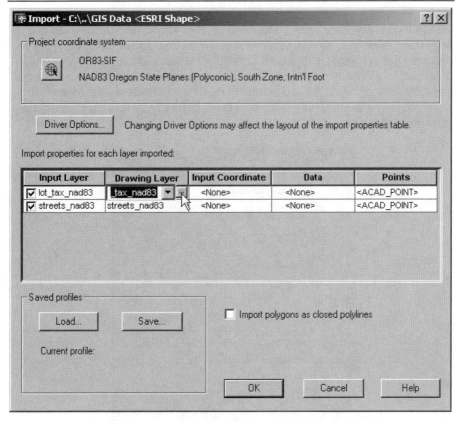

6. In the Import dialog box you can select any or all files in the current directory to Import into LDT. We will select both the **taxlot** and **streets** files.

7. Click on the **Drawing Layer** field in the **lot_tax_nad83** row to activate << ... >> button. Click on this button to bring up the **Layer Mapping** dialog box.

Here you can choose to import the drawing objects onto an existing layer, create a new layer, or select a column of data from the file that you are importing to determine the layer names. This last option will allow you to do some basic thematic mapping during the import of the objects. For example, if you were importing parcel data and that data set had a column for zoning. You could have the import command create a new layer for each zoning type and place each parcel on the appropriate layer for it's zoning designation.

8. For our exercise we will import the **lot_tax_nad83** ShapeFile onto a **new layer** called **"Tax Lot"**.

9. Click **<<OK>>**.

10. Repeat the process for the ShapeFile **streets_tax_nad83** setting it be **Created on a new layer** named **"Streets"**.

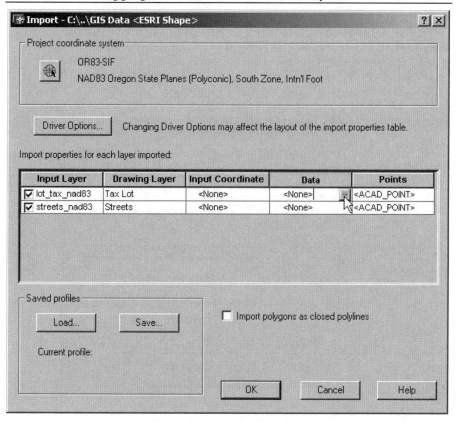

11. Back in the **Import** dialog box click on the **Data** field in the **lot_tax_nad83** row to activate **<< ... >>** button. Click on this button to bring up the **Attribute Data** dialog box.

Here we will create Object Data from the shape file's attribute data. You can enter the desired name for the Object Data Table and select the desired fields to import.

12. For our exercise we will import all fields and use the default name for the Object Data table.

13. Click <<OK>>.

14. Repeat assigning the data import options for the Streets coverage. But this time click the **<<Select Fields>>** button. This will allow you to select only the attribute data you wish to import rather than the entire database.

15. Deselect all fields except **"NAME_LONG"** and **"STREET_TYP"**.

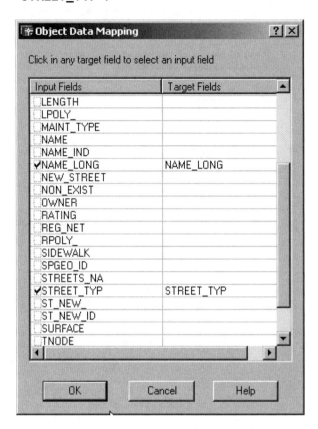

16. Click **<<OK>>** to return to the Import Dialog box. The completed dialog box should look like the one below.

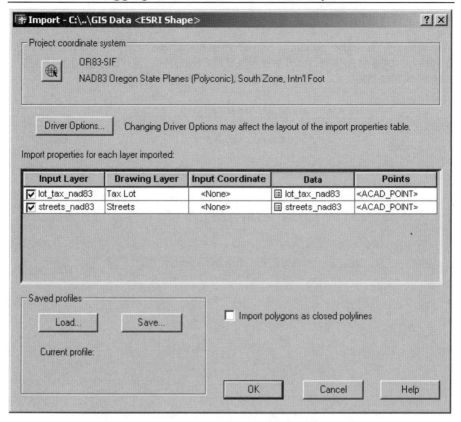

17. Click **<<OK>>** in the Import dialog box to begin importing the data.

18. **Zoom Extents** to view the imported data.

2.1.2 Controlling The Display Of Polygons

You will see that the polygons are all displayed with a solid hatch fill. This display option is a feature of the MPOLYGON object. To display just the edges of the polygons you need to set the polygon display mode.

1.   At the command line enter:

     ```
     Command: POLYDISPLAY
     ```

2.   Then use the option "**E**" to display only the edges of the polygons.

```
Enter mpolygon display mode [Edges only/Fill only/Both]
<Both>: E
```

3.   **Regen** to redisplay the Polygons.

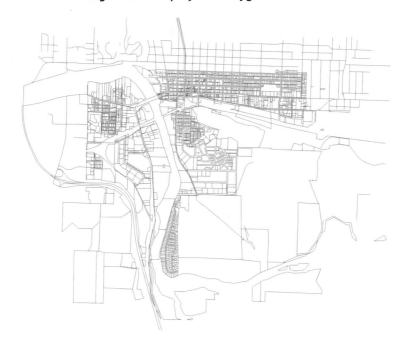

2.1.3 Viewing GIS attributes in AutoCAD

When talking about GIS data the term attributes refers to any type of attached data not just block attributes as we are used to in AutoCAD terminology. In Map attribute data can be object data, external database data, or block attributes. Within AutoCAD, the Map functions enable you to view and edit this attribute data, attributes imported from GIS files as well as attributes created as object data created in Map. To view the attributes imported above follow these steps.

Viewing imported Object Data:

1. Select **Map** ⇒ **Object Data** ⇒ **Edit Object Data.**

2. Then select the desired taxlot or street to display it's object data.

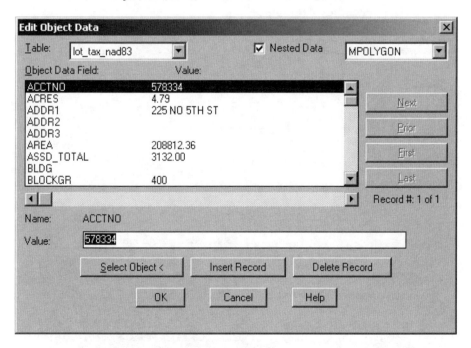

3. Here you can edit any object data as needed. Click <<**Cancel**>> to leave the **Edit Object Data** dialog box without making any changes.

2.2   Using Queries to Manage and Share Data

We will use Autodesk Map's ability to query information from other drawings to import, edit, and save back specific taxlot and street information. This gives us the ability to work with small, specific, pieces of larger data sets as well as allowing us to share the data with others that may need to be using it at the same time. We will also use queries in a later exercise to select spot elevation and breakline data for a specific area from a large citywide data set.

2.2.1   Creating A Preliminary Design Drawing

In order to use a query you must start a new drawing and attach the drawings that contain the data that you want to query.

1. So, start a new drawing named **"Preliminary Design"** using the same project **"Cadapult Level 1 Training"** and the **acad.dwt** template.

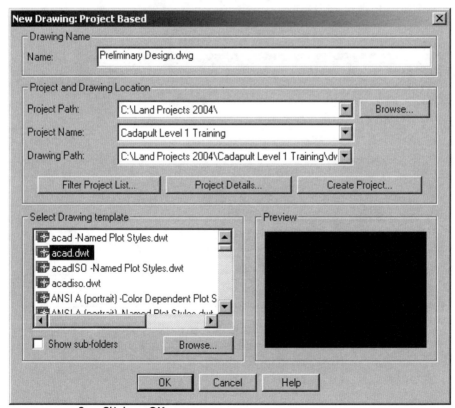

2. Click <<**OK**>>.

3. Save the changes to the Base Map drawing if prompted.

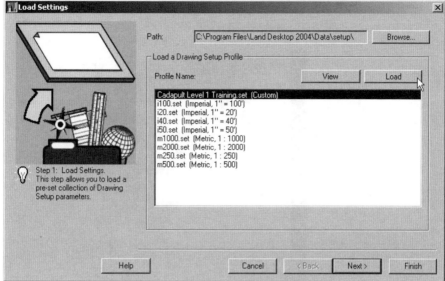

4. Load the **"Cadapult Level 1 Training"** Setup Profile when the first panel of the Drawing Setup wizard appears.

5. Then click **<<Finish>>** to skip the rest of the wizard.

6. Click **<<OK>>** to dismiss the **Finish** dialog box after reviewing your settings.

2.2.2 Attaching Source Drawings

1. In the Project Workspace, the large white area on the left of your screen, **right-click** on **Drawings** and select ⇒ **Attach** from the pop up menu.

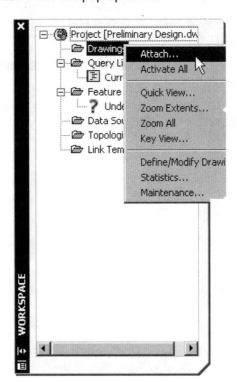

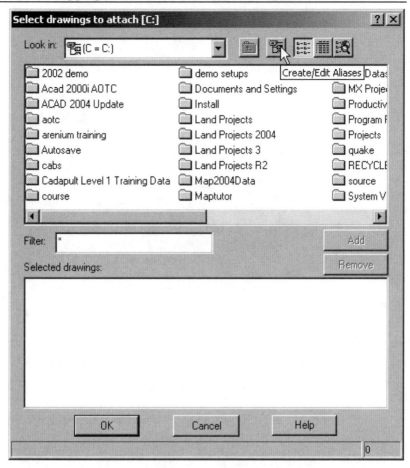

To attach drawings to Autodesk MAP you must have a drive alias. By default you have one for the C drive. A drive alias is a named shortcut to a drive or directory on your local machine or network. Its purpose is to aid in sharing data with others whose directory structure is different than yours. You can maintain the same drive alias and each have your own paths to the data.

> 2. Click the **second button from the left** at the top of the dialog box to **Create or Edit a drive alias.**

3. In the **Drive Alias Administration** dialog box enter **"TRAINING_PROJ_DWG"** for the alias name.

4. Use the **<<Browse>>** button to add the path.

5. For this exercise the path will be:

C:\Land Projects 2004\Cadapult Level 1 Training\dwg\.

6. Then click **<<Add>>** to add the alias. You will see it added to the list above.

If you click <<Close>> before you click the <<Add>> button, the alias will not be saved.

7. After you have added the alias, click **<<Close>>**.

Back in the **Select Drawing** dialog box you can choose the
TRAINING_PROJ_DWG alias from the drop-down list at the top of the
dialog box. Then you will see all of the drawings in that alias.

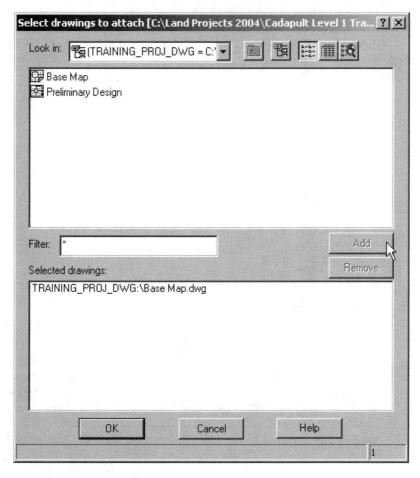

8. Select **Base Map** and click **<<Add>>**.

9. Then click **<<OK>>**.

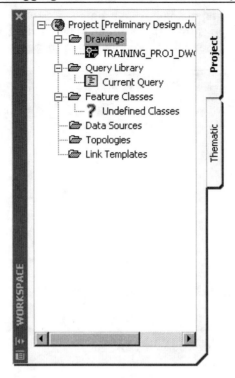

You will now see the drawing listed in the project workspace. The attached drawing is shown with a Blue icon in the project workspace. The Blue icon indicates that the drawing is active. Active drawings will be used for all Query and Quick View commands. Right-clicking on attached drawings in the project workspace and choosing deactivate can deactivate them. Deactivated drawings are still attached to the Map project but will not be used in **Queries** or **Quick Views.**

10. **Right-click** on **Drawings** and select ⇒ **Quick View.**

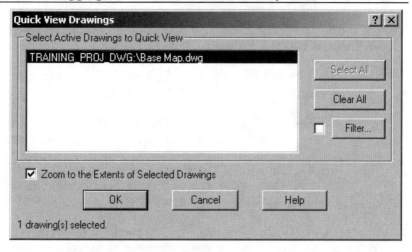

11. Confirm that the option to **Zoom to the Extents of Selected Drawings** is **Enabled**.

This will zoom your current project drawing to the extents of the attached drawings. If you don't do this the quick view image will appear in an area that is not shown on your screen.

A Quick View will show you a preview image of what is contained in any attached and active drawings. The quick view image will go away when you use a Redraw or during an AutoSave. If you lose the quick view just run the quick view command again to redisplay it.

2.2.3 Defining A Query

You can use any combination of 4 different Query Types to create your query. They are Location, Property, Data, and SQL.

- Location allows you to limit your Query to a specific area which you define within the source drawings.

- Property allows you to select the object to be queried by any AutoCAD property such as Layer, Linetype, Color, etc.

- The Data and SQL options allow you to select options for your Query according to their attached data; both object data and external database data.

By using the above Query types together you can limit the objects that are Queried into your drawing giving you a very specific selection set and keeping your drawing sizes smaller.

1. **Right-click** on **Current Query** in the workspace and select ⇒ **Define**.

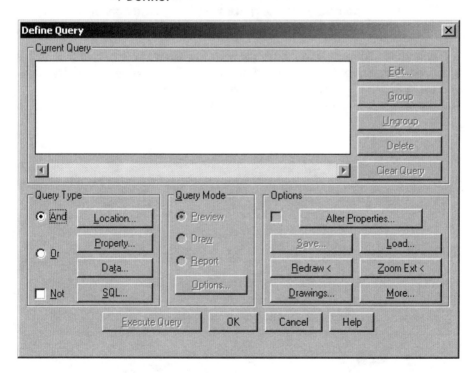

2. Click the **<<Location>>** button in the **Query Type** section.

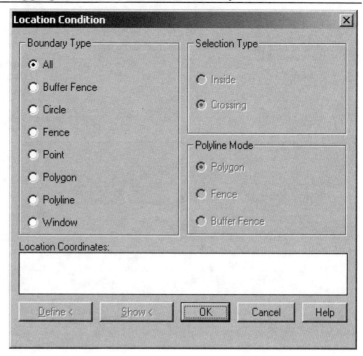

3. Choose a Boundary type of **All**.

4. Click **<<OK>>**.

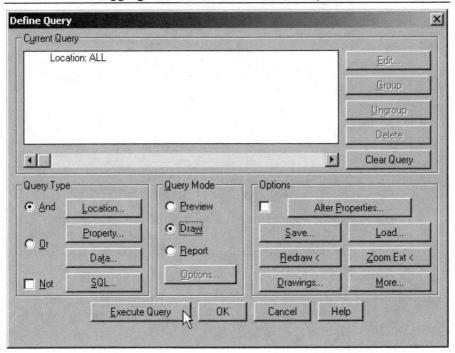

5. Now back in the **Define Query** dialog box; change the query mode to **Draw** and click **<<Execute Query>>**.

This will bring all of the objects in the attached drawing or drawings that meet the query criteria into the current Drawing. They are brought in as regular AutoCAD objects so you can edit them using any of your AutoCAD commands. Any changes can be saved back to the source drawings.

2.2.4 Saving Changes Back To The Source Drawings

Any changes made to a queried object can be saved back to the source drawing that the object was queried from. This is a very powerful feature but also one that you need to be very careful with. Using the Save Back command when it is not appropriate can edit base data and cause you to lose valuable information. Remember, the Undo command does not work with the Save Back command.

1. The queried polygons will be displayed as filled polygons. This can be difficult to edit so we should change the polygon display mode. At the command line enter:

   POLYDISPLAY

2. Then use the option **E** to display only the edges of the polygons.

3. **Regen** to redisplay the Polygons.

4. Edit any of the Queried objects using standard AutoCAD commands.

5. Once you finish the edit you will be asked if you want to add objects to the **Save Set**.

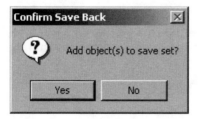

6. Click the **<<Yes>>** button to add the object or objects that you just edited to the save set.

This has not saved the edited objects back to the source drawing yet. It has only added them to a group of objects that can be saved back when you are ready.

7. Select **Map** ⇒ **Save Back** ⇒ **Save to Source Drawings**.

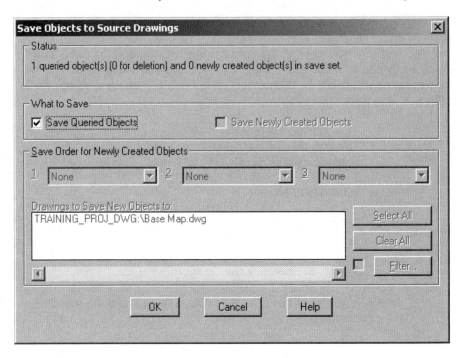

8. Click **<<OK>>** to save the object in the save set back to the source drawings.

You will notice that you object(s) that you saved back have disappeared from your drawing. To bring them back you can just run the query again.

9. Now **open** the drawing **BaseMap.dwg** to see that your changes were saved back to their source.

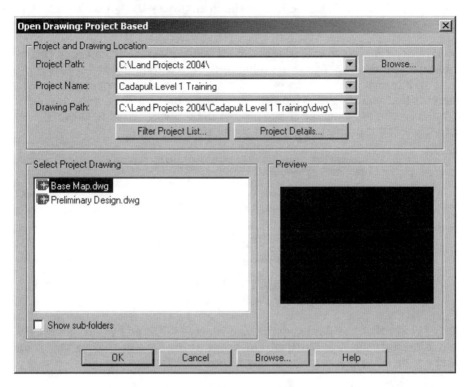

When you try to open a drawing that is attached and active in your current drawing you will receive the following error message.

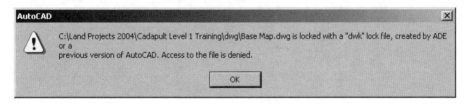

10. Click **<<OK>>** to clear the error message.

11. Click **<<Cancel>>** to close the Open Drawing dialog box.

12. **Right-click** on the **Base Map** drawing in the Project workspace and select ⇒ **Deactivate.** This removes the lock on the drawing but keeps the drawing attached for future use.

13. Now you can **Open** the **Base Map** drawing.

14. Save the changes to the **Preliminary Design** drawing when prompted.

A Save or an AutoSave will trigger the following message from Map. This is not an error, even though it has a big red circle with and X in it. This is just an informational message explaining that if you close the drawing the objects that you have queried will no longer know that they were queried from a source drawing. So if you reopen the drawing, edits to these objects will not prompt you to add them to the save set.

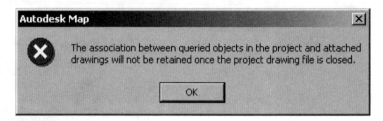

15. Now you can view the changes that you saved back from the Preliminary Design drawing.

Don't worry that the Base Map drawing is now corrupted by your edits. We made these changes to illustrate the usage of the Save Back command. We will not be using this drawing in any future exercises.

16. **Reopen** the drawing **Preliminary Design.dwg**.

17. **Right-click** on the **Base Map** drawing in the Project workspace and select ⇒ **Detach.**

2.3   Inserting Images And The Project Boundary

These exercises will look at ways to add raster images and a block containing the project boundary to our drawing.

2.3.1   Inserting Registered Image (Rectified Aerial Photography)

It is important to use the **Map** ⇒ **Image** ⇒ **Insert** command rather than the commands in the AutoCAD Insert pull-down because this command utilizes correlation information, in our example World Files (.tfw), to automatically place your image correctly in your coordinate system. Otherwise you would need to register each image manually.

    1.   Create a layer called **"Image"** and set it **Current**.

    2.   **Select Map** ⇒ **Image** ⇒ **Insert**.

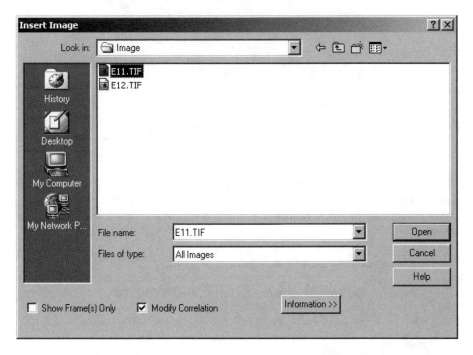

    3.   Browse to **C:\Cadapult Training Data\LDT Level 1 \Image**.

    4.   Select the file **E11.TIF**.

    5.   Click **<<Open>>**.

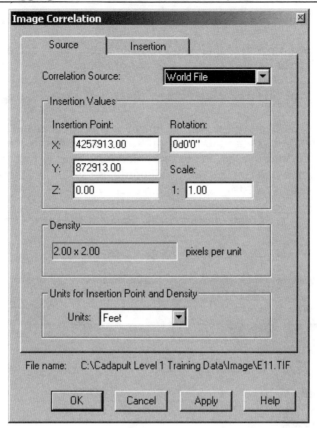

You will see that Land Desktop has automatically found the World File and set the Correlation Source and Insertion Values. You will also see a frame in your AutoCAD drawing editor showing the proposed location of the image.

     6. Click **<<OK>>** to inset the image.

     7. Repeat the process to insert image **E12.TIF**.

Images, by default, will cover everything in your drawing. To correct this "display order" problem and move the image back behind other drawing information follow these steps.

8. From the main menu select **Tools** ⇒ **Display Order** ⇒ **Send to Back**, and then click on the image boundary box (frame.) If the Tools pull-down is not available you will need to change menu palettes. To do this select **Projects** ⇒ **Menu Palettes**. Then load the **Autodesk Map 2004** Menu Palette.

The aerial photographs are useful background information however they do take a lot of system recourses to process and regenerate. So, when you are not using them it is best to turn them off.

9. Select the image by its frame and right-click. You must do this one image at a time.

10. From the popup menu select **Image** ⇒ **Show Image**.

11. Now only the frame of the image is shown and regenerated.

You may elect to " FREEZE" this layer as a valid alternative.

2.3.2   Adding the Project Area

We will now insert a block that contains a polyline defining our project area. This way we all are working with the same data.

1.   Select **Insert ⇒ Block.**

2.   Browse to the file:

C:\Cadapult Training Data\LDT Level 1\Drawings\Project Boundary.dwg

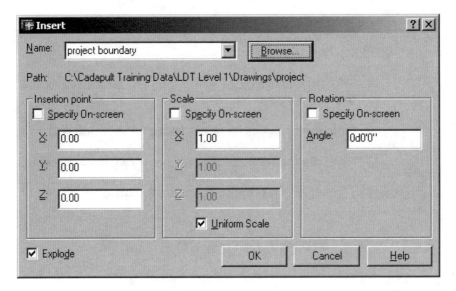

3.   If it isn't already, disable the to Specify the Insertion point on the screen.

4.   If it isn't already, enable the option to **Explode** the block.

5.   Click **<<OK>>** to insert the project boundary.

6.   Switch back to the **Land Desktop 2004** Menu Palette.

7.   Set layer **0 Current.**

8.   **Freeze** the **Tax Lot, Streets**, and **Image** layers.

2.4 Importing Point And Breakline Information From Aerial Mapping

We will use Autodesk Map's ability to query information from other drawings to bring in spot elevation and breakline data. The breakline and point data are stored in a tiled format to maintain smaller drawing sizes. Our project, like many others, does not fit neatly inside of one of these tiles, so we will bring data together from several drawings to fit our project area. Autodesk MAP allows you to attach multiple drawings and perform simple or very complex queries across tiled drawings.

2.4.1 Attach The Source Drawings

1. In the Project Workspace **right-click** on **Drawings** and select ⇒ **Attach**.

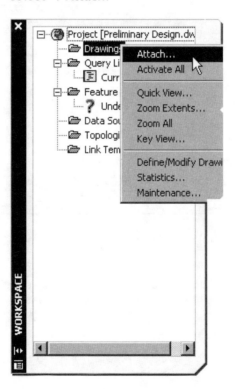

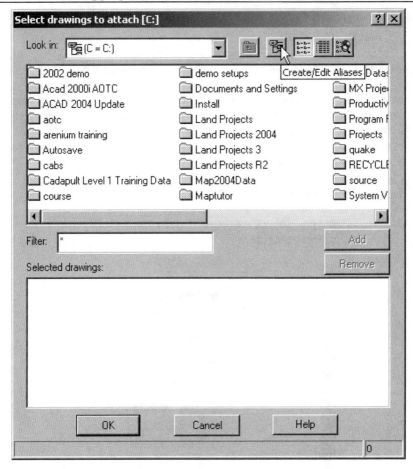

We will need to create another Drive Alias, the same way we did in our earlier exercise.

> 2. Click the **second button from the left** at the top of the dialog box to **Create or Edit a drive alias.**

3. In the **Drive Alias Administration** dialog box enter **"TRAINING_DATA"** for the alias name and click the **<<Browse>>** button to add the path.

4. For this alias the path will be:

C:\Cadapult Training Data\LDT Level 1\Drawings

5. Then click the **<<Add>>** button to add the alias. You will see it added to the list above.

If you select close before you click the add button the alias will not be saved.

6. After you have added the alias, click **<<Close>>**.

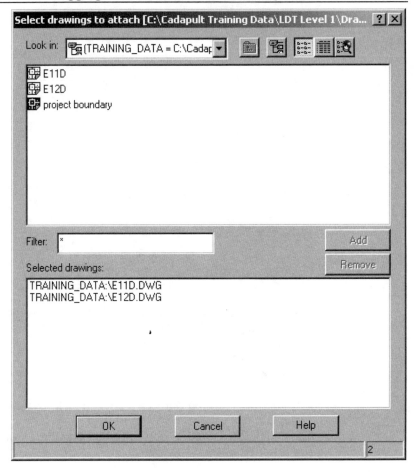

7. Back in the **Select Drawing** dialog box you select the **TRAINING_DATA** alias at the top of the dialog box. Then you will see all of the drawings in that alias.

8. Select **E11D** and **E12D** and click **<<Add>>**.

9. Then click **<<OK>>**.

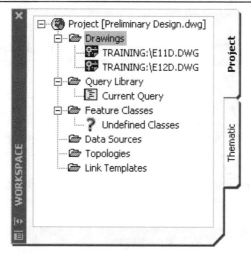

You will now see the two drawings listed in the project workspace.

10. Right-click on one of the drawings and select ⇒ **Quick View**. This will show you a preview image of what is contained in the attached drawings. The Quick View image will go away when you use a Redraw.

2.4.2   Defining A Compound Query

You can use any combination of 4 different Query Types to create your query. They are Location, Property, Data, and SQL. We discussed the details of each query type previously in section 2.2.3.

By using several Query types together you can limit the objects that are queried into your drawing giving you a very specific selection set and keeping your drawing sizes smaller. This is typically referred to as a Compound Query.

1. **Right-click** on **Current Query** in the workspace and click <<Define>>.

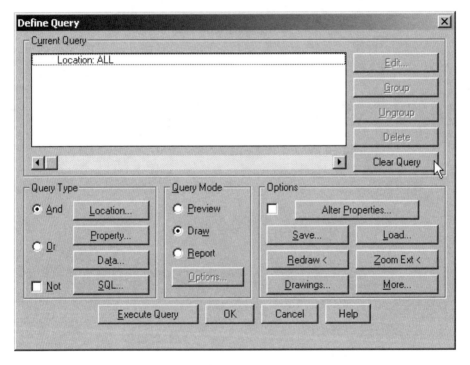

2. The previous query is still saved in the query dialog box. Click the <<**Clear Query**>> button to clear the previous query criteria and start a new one.

3. Click the <<**Location**>> button in the **"Query Type"** section.

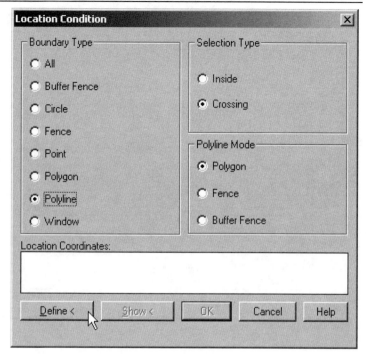

4. Choose a Boundary type of **Polyline**.

5. The Selection Type should be **Crossing**.

6. The Polyline Mode should be **Polygon**.

7. Click the **<<Define <>>** button.

8. When asked to select the polyline, pick the site boundary that is on the layer Project Boundary. You will see a preview of the buffer we created on the screen.

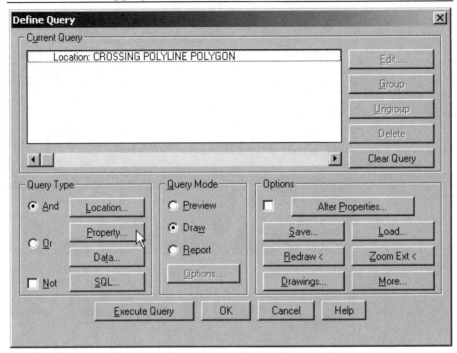

9. Now back in the **Define Query** dialog box, in the **Query Type** section click **<<Property>>**.

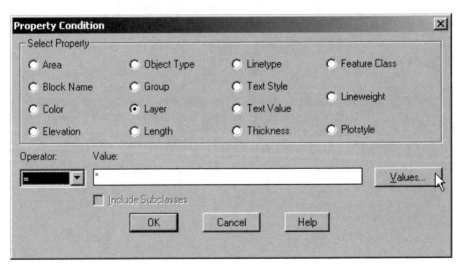

10. Set the property type to **Layer**.

11. Click the **<<Values>>** button.

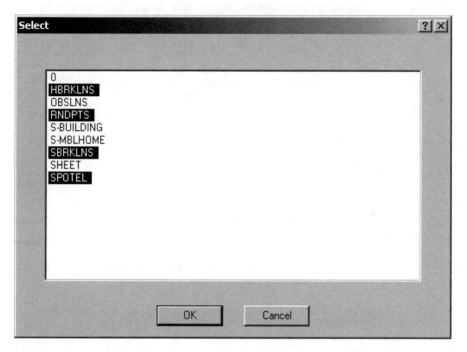

You are now shown a list of all of the Layers in all of the attached drawings that are active.

      12. Select the layers: **HBRKLNS, RNDPTS, SBRKLNS, SPOTEL.**

Hold down the control key to select more than one layer at a time.

      13. Click **<<OK>>**.

14. Click **<<OK>>** to save our changes to the Property Conditions dialog box.

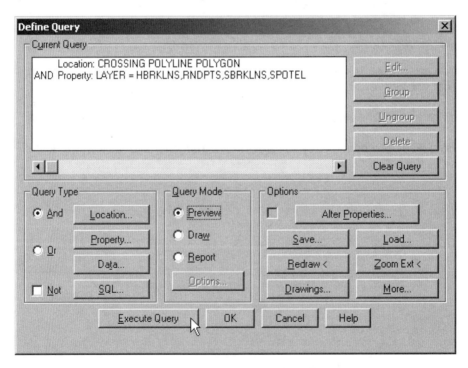

15. Set the Query mode to **Preview** and click **<<Execute Query>>**.

Using Preview mode will display a Quick View of the objects that meet the criteria you set in the Query. You can clear that display with a **Redraw**.

16. **Right-click** on the Current Query in the Project Workspace and select ⇒ **Define**.

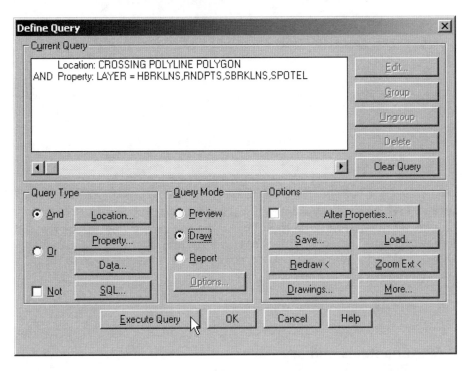

You will see that all the information from the previous query is retained.

17. Change the query mode to **Draw** and click **<<Execute Query>>**.

This will bring all of the objects that you saw in the Preview into the Drawing. They are brought in as regular AutoCAD objects so you can edit them using any of your AutoCAD commands. Any changes can be saved back to the source drawings. Be careful with **"Savebacks"**. In many situations, like ours here, you would never want to save changes back to the source data. That would change and corrupt the information from our aerial survey.

18. To prevent any changes from being saved to the source drawing **<<right-click>>** on each source drawing and select ⇒ **Detach**.

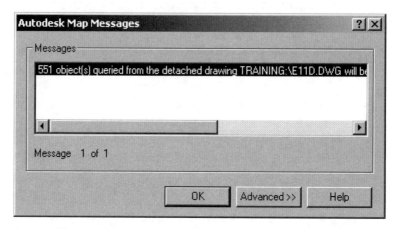

When you detach the drawings you will get a message that says that the objects we queried will lose their link to the source drawing and be treated as newly created objects. This is not an error message; it is exactly what we wanted to do. We have lost the ability to save changes back to the source drawing.

19. **Save** the drawing.

2.5   Chapter Summary

In this chapter we have used Land Desktop, and the Map components of Land Desktop in particular, to collect base data for our project from a variety of sources. It is important to keep in mind that every project is different and an important first step is to think about what type of data that you have available and what you want to collect. You can then use a variety of different commands, as they are appropriate, to compile that data similar to the way we have in this chapter.

Chapter 3

Preliminary Layout

In this chapter we will build a preliminary existing ground surface from the aerial survey data we collected in the previous chapter, layout and define two preliminary alignments, work with the point settings, and export preliminary design points to the surveyors. You will be introduced to many different Land Desktop commands in this chapter. But don't worry if you still have a few questions about them because we will cover them again in more detail in later chapters. This is meant to introduce you to the topics and provide repetition with different examples later.

- **Creating a Preliminary Existing Ground Surface**

- **Creating Preliminary Alignments**

- **Creating Points From An Alignment**

- **Creating A Point Group And Exporting The Points For Field Verification**

Dataset:

To start this chapter we will continue working in the drawing named **Preliminary Design.dwg**. You can continue with the drawing and project data that you currently have from the end of the previous chapter or you can install the dataset named **"Chapter 3 Cadapult Level 1 Training.zip"** from the CD that came with the book. Extract the dataset to the folder **C:\Land Projects 2004** or whatever folder you have been using as your project path. Extracting the project from the dataset provided will ensure that you have the project and drawings set up correctly for the exercises in the following chapter and overwrite any mistakes that you may have made in previous exercises.

3.1 Creating a Preliminary Existing Ground Surface

This set of exercises will look at creating a preliminary surface of the existing ground from the AutoCAD objects that we queried into the drawing in the previous chapter. In later chapters we will build a surface from survey data and merge it with the preliminary surface using the preliminary surface to add buffer data around our survey.

3.1.1 Setting the Surface Display Settings

Similar to most of the features in Land Desktop, Surfaces and their display are controlled by the Drawing Settings. These settings are originally copied from the project prototype when you create a new drawing. Understanding these settings and configuring them before you start working on their associated feature in the drawing will make your work in LDT much easier.

1. If it isn't already, open the drawing **Preliminary Design.dwg** from the project **Cadapult Level 1 Training**.

2. Select **Projects ⇒ Edit Drawing Settings**.

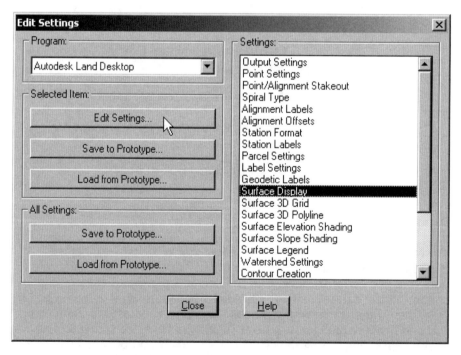

3. Select **Surface Display** from the list and click **<<Edit Settings>>**.

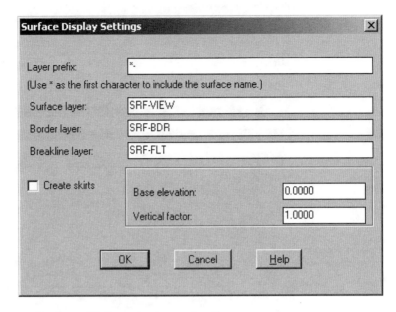

4. Enter an "*-" as the Layer Prefix.

This will use our surface name as a prefix for all of our surface layers.

5. Click <<OK>> and <<Close>> to exit all dialog boxes.

Any of the settings found in the Edit Settings dialog box can be saved to a project prototype by selecting the Save To Prototype option. By doing this these settings will be used as defaults every time you create a new drawing in a project the uses that prototype.

3.1.2   Adding Surface Data

1. Isolate the spot elevation layers:

RNDPTS
SPOTEL

You can do this by typing **LAI** at the **command line**. Then either select objects on the layers you would like to isolate or **ENTER** to bring up a dialog box with a list of layers to select from.

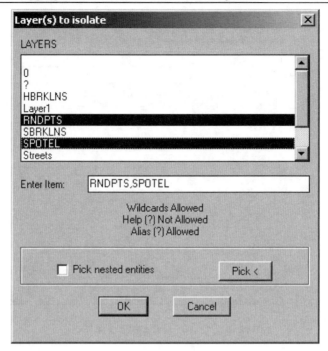

2. Select **Terrain** ⇒ **Terrain Model Explorer**.

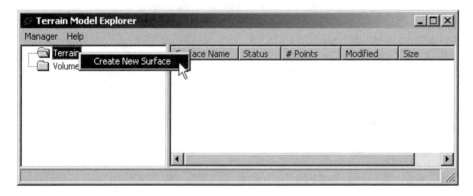

3. **Right-click** on the **"Terrain Folder"** and select
   ⇒ **Create New Surface**.

This creates a new folder in our project, which will hold the information to build the new surface. The new surface is automatically named **"Surface 1"**.

4. Right-click on **Surface 1** and pick ⇒ **Rename**. Rename the surface **"Pre-EG"**.

5. Expand **"Pre-EG"** by double clicking on it.

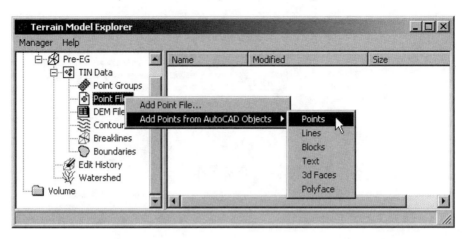

6. **Right-click** on **Point Files** and select **Add Points from AutoCAD Objects** ⇒ **Points**.

7. Select Objects by `Entity`.

8. Select all the points with a crossing window.

This creates a point file with the x,y,z coordinates of all the points that you selected. This point file will be used to build our surface.

9. Isolate the breakline layers:

    HBRKLNS
    SBRKLNS

10. In the terrain model explorer right click on **Breaklines** and select ⇒ **Define by Polyline**.

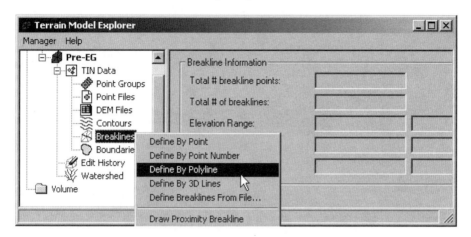

11. When asked for **breakline desciptions**, hit **ENTER** for none. This will give the breaklines all the description of **"Unclassified"**.

12. Select all of the breaklines with a **crossing window**.

13. Choose **Yes** when asked to delete the existing objects.

14. Turn on the layer **"Project Boundary"**.

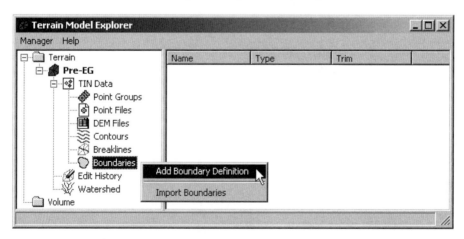

15. In the terrain model explorer right-click on **Boundaries** and select ⇒ **Add boundary definition**.

16. Select the polyline that we will use as the boundary of our surface.

17. When asked for a boundary name return to use the name **Boundary0**.

18. The boundary type will be **Outer**.

19. Answer **No** when asked if you want to make breaklines along the boundary.

20. **ENTER** when asked to select (another) Polyline because we will only be using one boundary for this surface.

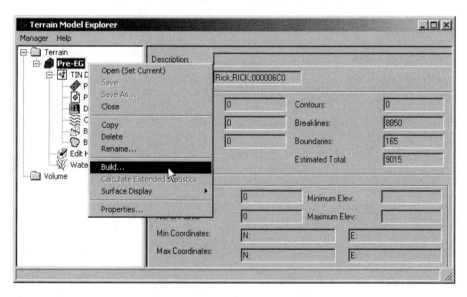

21. In the terrain model explorer **right click** on the surface name **Pre-EG** and select ⇒ **Build**.

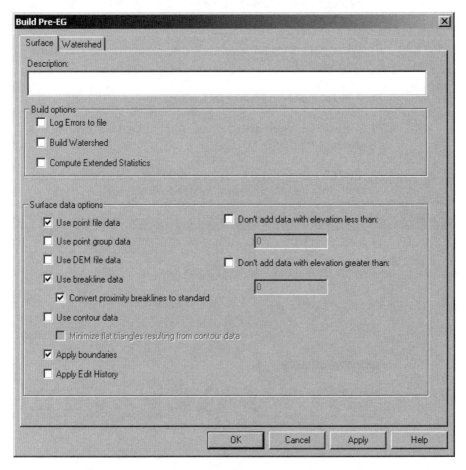

22. In the Build Surface dialog box make sure that you have the following options checked: **Use point file data, Use breakline data**, and **Apply boundaries.**

23. Click **<<OK>>** to build the surface.

3.1.3 Displaying and Reviewing the Surface

Once a surface is built there are a number of ways to view it. In this exercise we will view the surface as a Polyface Mesh. A Polyface Mesh is a group of 3D faces combined into a single object. This gives us one singe object to control rather than many 3D faces or 3D lines. And since it is made up of 3D faces we can shade it once we rotate the view.

1. Back in the Terrain Model Explorer **right-click** on the surface name **Pre-EG** and select ⇒ **Surface Display** ⇒ **Polyface Mesh.**

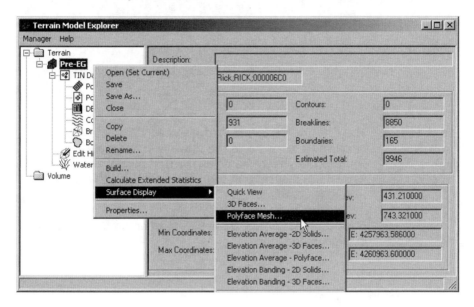

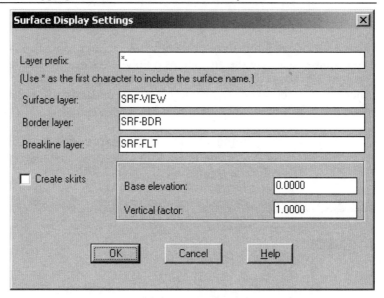

2. The **"*-"** for the **Layer Prefix** should already be entered because we added it to the drawing settings earlier. This will use the surface name as the first part of the layer name for all layers that are associated with this surface including contours.

3. **Erase** the old Surface View Layers when prompted at the command line.

To check the integrity of the surface from any angle we will use the Object Viewer.

4. Select the **Polyface Mesh**.

5. **Right-click** and pick ⇒ **Object Viewer** from the popup menu.

The Object Viewer window is now launched and you can rotate the surface around to any desired angle by picking and dragging your cursor. You can also **right-click** in the object viewer to select a shading mode.

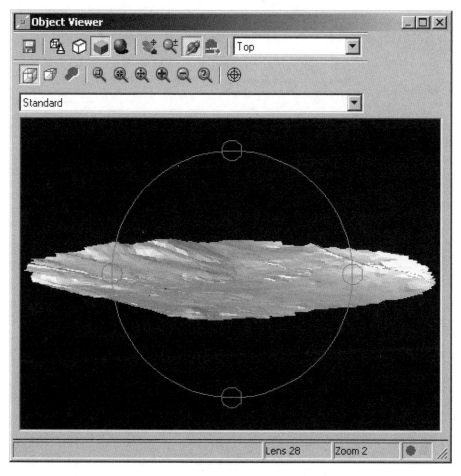

When finished examining the surface you can close the Object Viewer.

6. When you are finished viewing the surface you can erase the Polyface Mesh by selecting **Terrain** ⇒ **Terrain Layers** ⇒ **Surface Layer**.

7. Then at the command line, **ENTER** to select the default option of `Erase`.

The surface is stored in the Project so there is no reason to keep the objects in your drawing. They only take up space.

3.1.4 Quick Sections

Viewing a Quick Section is another good way to review a surface. The Quick Section lets you view a dynamic section through one or more surfaces.

1. Create a new **Layer** called **Section Lines** and set it **Current**.

2. Draw a polyline across the surface where you would like to view the section. This polyline can contain multiple segments and even arcs.

3. Select **Terrain** ⇒ **Sections** ⇒ **View Quick Section**.

4. Now pick the polyline that you drew earlier.

5. This will display the Section Viewer.

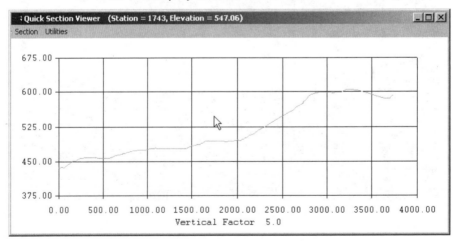

6. Now the fun begins because as you move or stretch the section line the Quick Section Viewer will automatically update.

7. You can even change the vertical exaggeration, labeling, and colors by selecting **Section ⇒ View Properties** in the Quick Section Viewer.

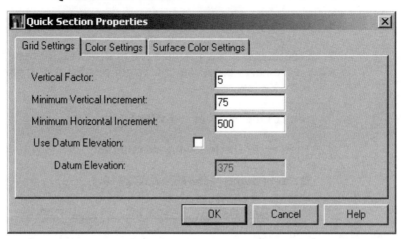

As an optional step you can draft the section into the drawing by selecting Utilities ⇒ Import Quick Section. The display of the drafted Quick Section can be controlled with layers and don't forget that it is only dynamic in the Section Viewer. So if you want to reposition your section line you will need to import the section to the drawing again. A grid can be added to your drafted section by selecting Terrain ⇒ Sections ⇒ Grid For Sections.

3.2 Creating Preliminary Alignments

Next we will layout a preliminary version of our horizontal alignment for the two roads in our project. This will be based on our aerial survey data and will need to be checked, and most likely modified, by the survey crew when they go on site and conduct their survey.

The alignment may be sketched in using the AutoCAD line, arc, and fillet commands if you wish. Land Desktop also provides a variety of COGO commands in the Lines/Curves pull-down. The result is the same whichever command you decide to use. You will be producing regular AutoCAD Line and Arc entities. These will later be defined into the Land Desktop project database as an alignment.

3.2.1 Drafting The Preliminary Horizontal Alignment

Layout "C Street":

1. Isolate the Project Boundary layer.

2. Create a **new layer** named Pre C Street CL for the preliminary centerline of C Street and set it **current** and color **Yellow.**

3. Select **Lines/Curves** ⇒ **By Direction.**

4. Then enter the following information:

   Starting point: **4258751.3,872710.1**

Quadrant	Bearing	Distance
2	88.2435	580
2	88.0505	435
2	88.1755	510

5. Now when asked for a Quadrant, hit **Enter** to end the line, and **Enter** again to end the command.

Layout "T Street":

6. Create a **new layer** named **Pre T Street CL** for the preliminary centerline of T Street and set it current and color Yellow.

7. Select **Lines/Curves** ⇒ **By Direction.**

8. Then enter the information from the following table.

   Starting point: **4258634.5,873277.6**

Quadrant	Bearing	Distance
2	83.5930	125
2	89.2525	1425
2	1.123	325

9. When asked for a Quadrant, hit **Enter** to end the line, and again to end the command.

10. To finish the T Street alignment use the AutoCAD **Line** command with an **End Point** osnap to continue where you just left off on the T Street alignment and a **Perpendicular** osnap to create the intersection with C Street.

11. Return to end the command and the Line.

Add horizontal curves to the alignment:

12. Select **Lines/Curves ⇒ Curve Between Two Lines.**

13. Pick the first two tangents (line segments) in the "T Street" alignment. Then enter the desired parameters for the curve. Use the following tables of curve information for the two alignments.

T Street

Curve #	Type	Value
1	Length	150
2	Radius	200
3	Tangent	50

C Street

Curve #	Type	Value
1	Length	250
2	Tangent	100

3.2.2 Defining the Alignment

Once the geometry of the alignments is laid out it needs to be defined into the LDT Alignment database.   Before you define the alignment into the Land Desktop database you can edit the alignment graphically using AutoCAD or LDT commands as you wish. Once you have defined the alignment into the LDT Alignment database you can continue to edit the alignment graphically using AutoCAD or LDT commands, however, these changes will not automatically update the project database. You will have to redefine the alignment and overwrite it in the project if you want to incorporate any graphical changes.

1. First isolate the centerline layers.

    Pre C Street CL
    Pre T Street CL

2. Select **Alignments ⇒ Define from Objects.**

3. Select the beginning tangent in the T Street alignment near the northwest end. This will be the starting point of the alignment.

4. Then pick the remaining tangents in the T Street alignment with a crossing window.

Beginning of T Street Alignment

This alignment will not have a reference point so . . .

5. **<<Enter>>** for the start of the alignment.

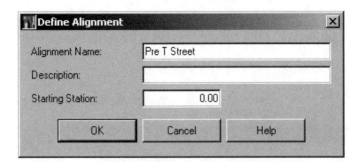

6. Enter the Alignment name of **"Pre T Street"**.

7. Repeat Steps 2 through 6 to define the **"Pre C Street"** alignment.

3.3 Creating Points From An Alignment

Now we will place horizontal points along the alignments and export these points to an ASCII file for our surveyors to verify in the field.

3.3.2 Establishing the Point Settings

1. Check the Point Settings, Select **Points** ⇒ **Point Settings.**

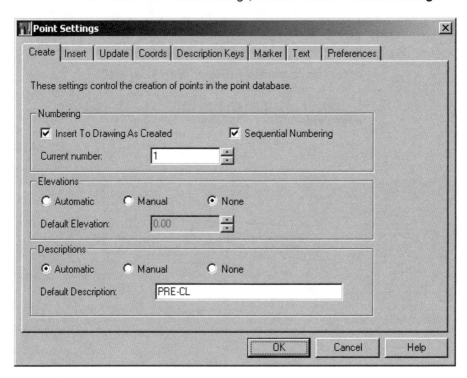

2. Enable **Sequential Numbering.**

3. Set the Elevations to **None.**

4. Set Descriptions to Automatic with a default description to **"PRE-CL".**

5. Select the **Insert** tab.

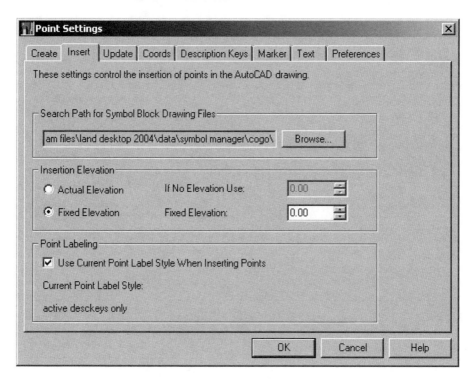

6. Use a fixed elevation of "**0**".

7. Confirm that the option to use the **Current Point Label Style When Inserting Points** is **enabled**.

If Point Labeling is not turned on then your description keys will not work.

8. Select the **Update** tab.

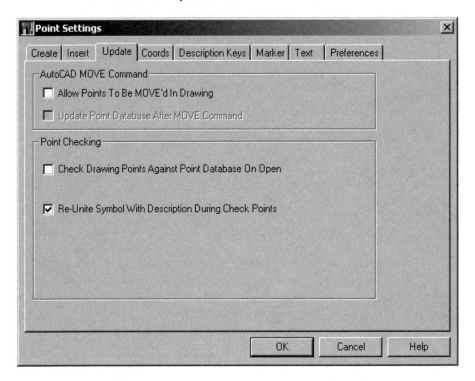

9. **Disable** the two options associated with the **AutoCAD Move**
   command.

With these settings if you use an AutoCAD Move or stretch command on
your points only the point label will be moved the actual location of the
point will remain fixed in the drawing.

10. Select the **Coords** tab.

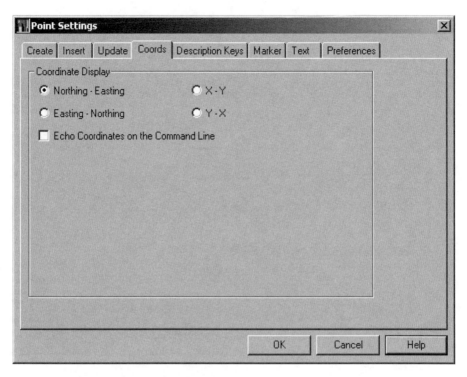

The **Coords** tab controls the display of the Point Coordinates at command prompts and during list operations in this drawing.

11. Select the **Description Keys** tab.

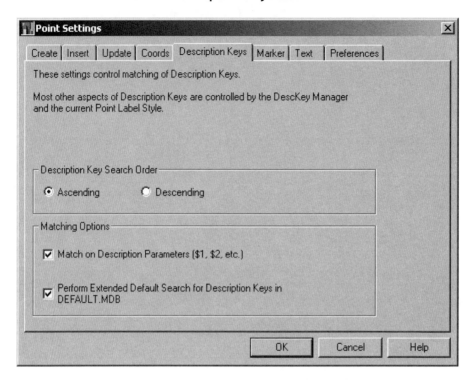

The **Description Keys** tab controls how your description file is sorted and searched.

12. Select the **Marker** tab.

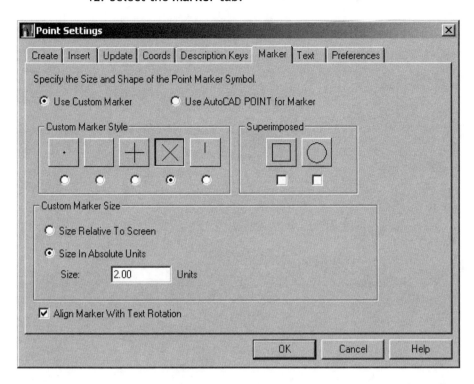

The **Marker** tab controls the display of your point marker and its size. This sets the default display settings for your point markers and controls the display of points as they are inserted into the drawing. If you want to change the display of points that are already in your drawing you must use the Display properties command: Points ⇒ Edit points ⇒ Display properties.

13. Set the **Absolute** size of the point marker to "**2**".

14. Select the **Text** tab.

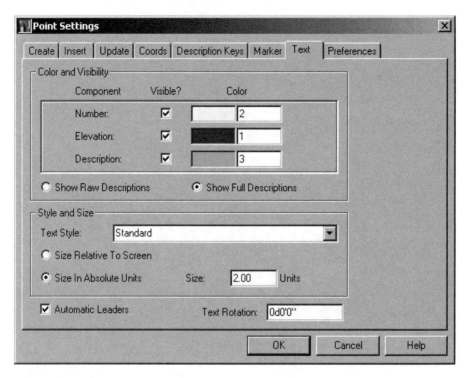

The **Text** tab will allow you to control the display of the text on your points as they are inserted into the drawing. Similar to the marker tab these are default settings for the points as they are inserted to the drawing. If you want to edit points currently in your drawings you will have to use the Display Properties command.

15. Set the text size to an **Absolute** size of **"2"**.

16. Enable the option for **Automatic Leaders**.

17. Select the **Preferences** tab.

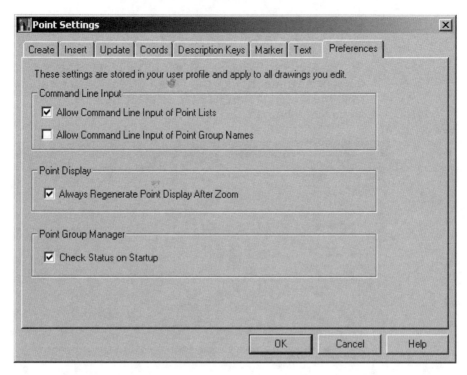

The preferences tab allows you to choose if you want to input point lists and point group names at the command line or in the dialog box. It also allows you to control if the point display is automatically regenerated after all zoom commands.

18. **Disable** the option to **Allow Command Line Input of Point Group Names.**

19. **Enable** the option to **Always Regenerate Points Display After Zoom.**

This is more important if you are using relative point sizes. Without this option the points may not always resize correctly.

20. Confirm that the Preferences Tab is setup as shown above.

21. Click **<<OK>>** to save the changes.

3.3.3 Setting Points on the Alignment

1. Create a new layer named **PNTS-PRE-CL** for the preliminary centerline points and set it current.

2. Set the Current alignment. Select **Alignments ⇒ Set Current Alignment.** Pick the **"Pre T Street"** alignment from the screen or hit **ENTER** to select it from the Alignment Librarian.

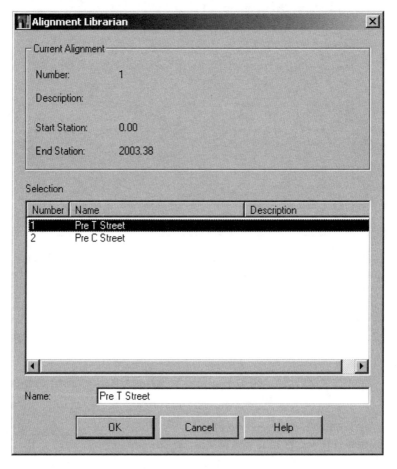

3. Select **Pre T Street** from the list.

4. Click **<<OK>>** to set Pre T Street as the current alignment.

Set the Preliminary alignment points.

5. Select **Points** ⇒ **Create Points - Alignments** ⇒ **Measure Alignment.**

6. Accept the default beginning and ending stations.

7. Use an offset of **"0"**.

8. Use a station interval of **"50"**.

9. Set the current point number to **"1"**. Points are now created at a 50' interval on the centerline.

10. Select **Points** ⇒ **Create Points - Alignments** ⇒ **At PC, PT, SC, etc...**

11. Accept the default beginning and ending stations and the current point number.

12. Repeat steps 2 through 11 to set points on the "Pre C Street" alignment.

3.4 Creating A Point Group And Exporting The Points For Field Verification

Create a Point Group of the Preliminary Centerline Points. This point group will be used to export these specific points to the surveyors.

3.4.2 Creating a Point Group

1. Select **Points** ⇒ **Point Management** ⇒ **Point Group Manager.**

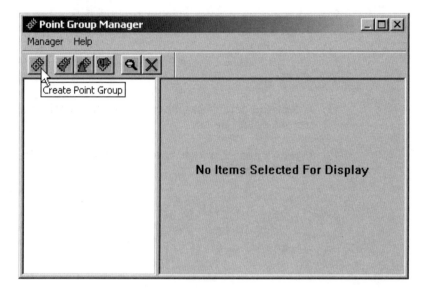

2. Click on the top left button **<<Create Point Group>>**.

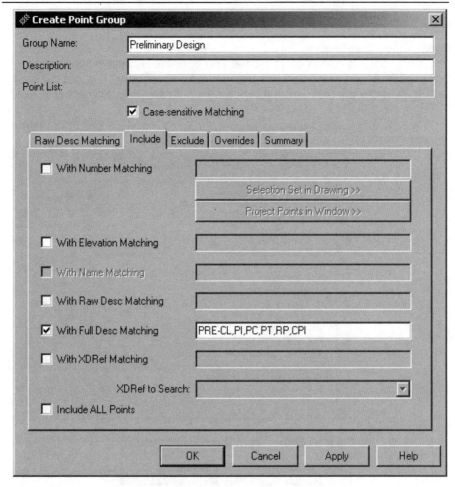

3. Create a new point group named **"Preliminary Design"**

4. Include Points with the Full Description Matching **PRE-CL, PI, PC, PT, RP, and CPI.**

5. Click **<<OK>>** to save the new Point Group.

6. **Close** the Point Group Manager.

3.4.2   Creating A Point Import/Export Format

First we will create an Import/Export format. Land Desktop comes with
many standard import/export formats predefined. However, you may at
some point need to define a new format for a specific project need. So we
will look at the process of creating your own custom import/export format.

1.   Select **Points** ⇒ **Import/Export Points** ⇒ **Format Manager**.

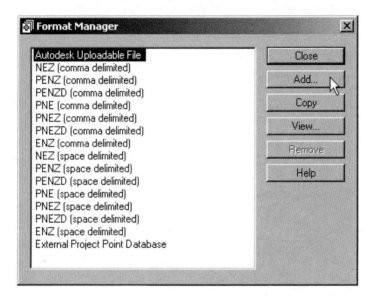

2.   Click **<<Add>>** from the Format Manager.

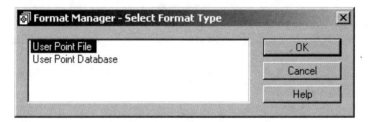

3.   We will create a format that uses a **User Point File**.

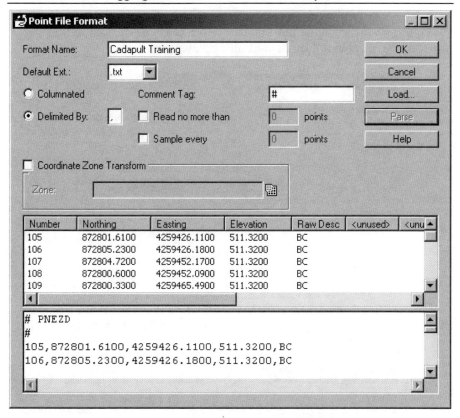

4. Name the format **"Cadapult Training"**.

5. Set the format to be **delimited by** a **comma**.

6. Enter the **"#"** symbol as a **Comment Tag**.

Land Desktop will ignore any line that begins with the Comment Tag symbol during the import of the file. So if you want to have information in the header of the point file, as we do in the example above, you must use a Comment Tag.

7. When picking any of the column headers at the bottom of the dialog box you can select the type of data in that column, by default all display as **unused.**

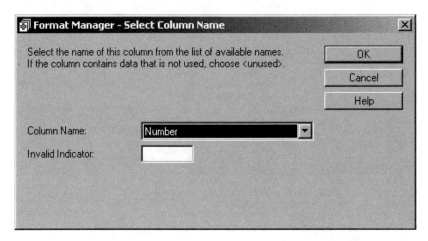

8. Create a format that contains columns for **Point Number, Northing, Easting, Elevation,** and **Raw Description** in that order.

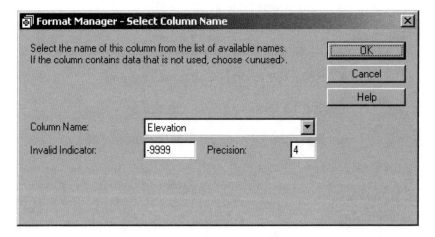

You can also use the Invalid Indicator option to set the value that your data collector uses for null entries.

9. When setting up the **Elevation** field set the **Invalid Indicator** to **-9999.**

10. To test your new format you can **<<Load>>** an example file that would use the new format. The example above has loaded the file:
C:\Cadapult Training Data\LDT Level 1\Points\Site.txt

Once you load the example file it will be displayed in the text box at the bottom of the dialog box.

11. With an example file loaded you can click **<<Parse>>** to test the new format.

12. Click **<<OK>>** when you are finished setting up the format to save it and exit the dialog box.

13. **<<Close>>** the main **Format Manager** dialog box.

3.4.3 Export The Points To An ASCII File

1. Select **Points ⇒ Import/Export Points ⇒ Export Points.**

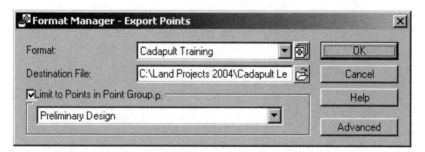

2. Set the Format to **"Cadapult Training".**

3. Set the destination File to:
C:\Land Projects 2004\Cadapult Level 1 Training\survey\Pre-cl.txt

4. Limit to points in the Point Group **"Preliminary Design".**

If you do not limit to the points in a point group then Land Desktop will export all the points in the project.

3.5   Chapter Summary

In this chapter we have used Land Desktop to create a surface from
AutoCAD objects. This is a process that you will often use if you are
receiving surface data from a photogrammetrist, surveyor, or anyone else
that doesn't happen to be using Land Desktop. Many programs now have
the option to import and export surface and other project data through the
Land XML format. However, there are still a large number of people using
software that does not support Land XML. In that case exchanging the
surface data as drawing objects is one of your only options. Which is why
this is an important process.

We also laid out a preliminary alignment and created points along it for
field verification. This was our introduction to the alignment commands,
which we will use again in detail later.

Chapter 4

Creating a Survey Plan

In this chapter we will work with description keys and point groups to organize our drawing and our project as we import survey data. We will also work with several of Land Desktop's COGO and labeling commands. The chapter finishes by working with parcels and the parcels sizing commands.

- **Create a Survey Drawing**

- **Description Keys**

- **Importing Survey Points**

- **Working With Point Groups**

- **Working With Parcels**

Dataset:

To start this chapter we will be creating a new drawing. You can continue with the project data that you currently have from the end of the previous chapter or you can install the dataset named **"Chapter 4 Cadapult Level 1 Training.zip"** from the CD that came with the book. Extract the dataset to the folder **C:\Land Projects 2004** or whatever folder you have been using as your project path. Extracting the project from the dataset provided will ensure that you have the project and drawings set up correctly for the exercises in the following chapter and overwrite any mistakes that you may have made in previous exercises.

4.1 Create a Survey Drawing

We will now create a new drawing in our project to display our survey data as we import it.

1. Select **File** ⇒ **New**.

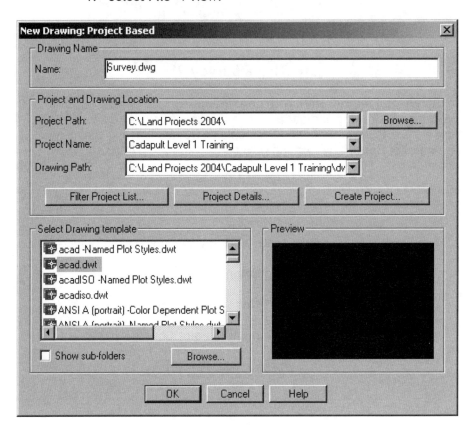

2. Enter **"Survey"** for the Drawing Name.

3. The Project information should still be set to **"Cadapult Level 1 Training"**.

4. Select **"acad.dwt"** as the Drawing Template.

5. Click **<<OK>>**.

6. Load the **"Cadapult Level 1 Training"** Setup Profile when the first panel of the Drawing Setup Wizard appears.

7. Then click **<<Finish>>** to skip the rest of the Wizard.

4.2   Description Keys

Description Keys control points as they are inserted into the drawing. They act as a filter between the project and the drawing and control the layer points are inserted on, the full description displayed on the point, and optionally insert a symbol with the point as well as place it on a specified layer. You are allowed to have more than one description key file in a project and point label styles are used to determine which description key file is used.

Definitions of the terminology used to create Description Keys:

DescKey Code:	Raw Description or the Description Entered in the field by the surveyors.
Description Format:	Full Description or the Description that will be shown on the Point Object in the drawing.
Point Layer:	Layer in the drawing that the point Object will be inserted on, regardless of what the currant layer is.
Symbol Block Name:	This field is optional. Use it if you want to insert a block with your point, for features like trees or fire hydrants.
Symbol Layer:	Controls the layer on which the above symbol will be inserted.

4.2.1 Creating a Description Key File

1. Select **Points** ⇒ **Point Management** ⇒ **Description Key Manager**.

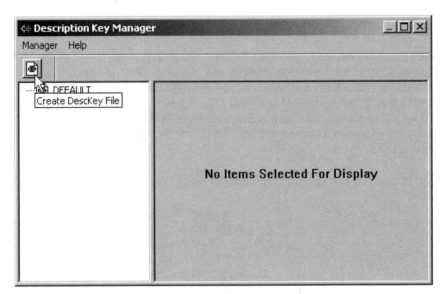

2. Click the white button on the left of the toolbar to create a new **description key file**.

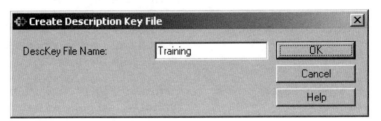

3. Name the new file **"Training"**.

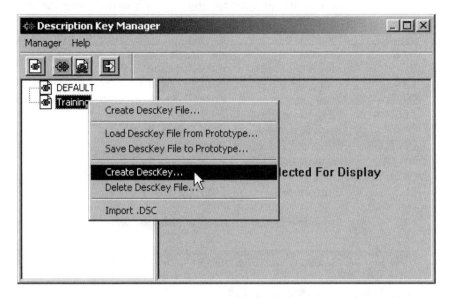

4. Now **right-click** on **Training** and select ⇒ **Create DescKey**.

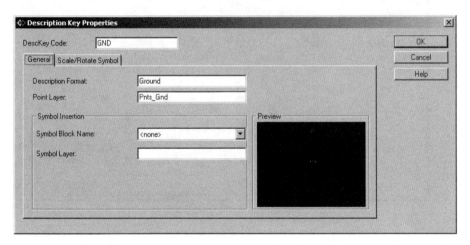

5. Enter a Code of **"GND"**.

6. The Description Format is **"Ground"**.

7. The Point Layer is **"Pnts_Gnd"**.

8. Click **<<OK>>** to save the new Description Key.

9. Repeat the process for the keys in the following table:

Code	Description	Layer	Symbol	Symbol Layer
GND	Ground	Pnts_Gnd		
BC	Building Corner	Pnts_Building		
CRNR	Prop_Cor	Pnts_Boundary		
DT	Tree	Pnts_Tree	cg_t15	Sym_Tree
AEC	Edge of Asphalt	Pnts_Aec		
DWYRK	Rock Driveway	Pnts_Driveway		
DWYCON	Concrete Driveway	Pnts_Driveway		
DWYAC	Asphalt Driveway	Pnts_Driveway		
LP	Light Pole	Pnts_Utiltity	lumin	Sym_Utility
CL	Center Line	Pnts_CL		

10. Close the **Description Key Manager** when you are finished.

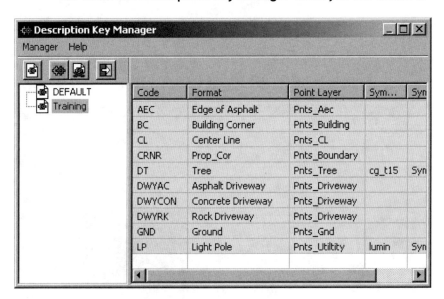

4.2.2 Point Label Styles

The Point Label Style controls the use of Description Keys.

1. Select **Labels** ⇒ **Show Dialog Bar**.

2. Select the **Point** tab.

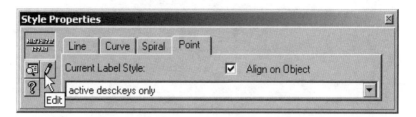

3. Click the **pencil icon** to edit the **"active desckeys only"** point label style.

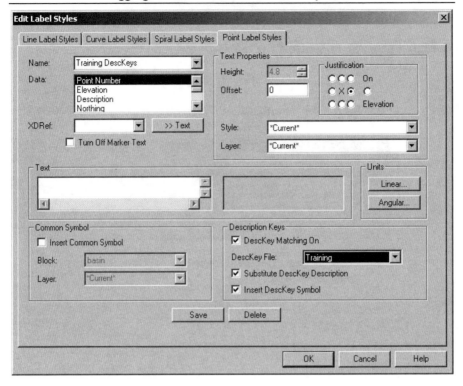

4. Enter a new style name called **"Training DescKeys"**.

5. In the lower right corner select **"Training"** as the **DescKey File**.

6. Click **<<Save>>**.

7. Click **<<OK>>**.

This has created a Point Label style that will automatically use our "Training" Description Key file whenever points are inserted to the drawing while it is current.

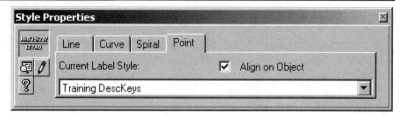

8. Set the new style you created, **"Training DescKeys"**, as the current Point Label Style.

The Style Properties dialog box is a modeless dialog box. That means it can stay open on your screen as long as you like. You can continue and use other commands while it is open without causing any problems.

9. Create a **new layer** named **Pnts-Misc** for miscellaneous points that may be missed by your description keys and set that layer **current**. After the points are inserted into the drawing you can isolate this layer to see if there are any points on that were missed. If there were it would indicate either a problem with a description key or a miscoded point.

4.3   Importing Survey Points

Now that the description keys are set up and ready to use you are ready to import the survey points into the project. The import command will not only import the point into the project but it will also insert the points into the drawing.

4.3.1   Importing Points From An ASCII File

It is important to understand the difference between the **Import Points** command and the **Insert Points to Drawing** command. The Import Points command populates the project point database and also inserts points into the drawing. This is a command that you will typically use only once per point file. While you may use the Insert Points to Drawing command many times to bring points that already exist in the project into the drawing.

    1.   **Select Points ⇒ Import/Export Points ⇒ Import Points.**

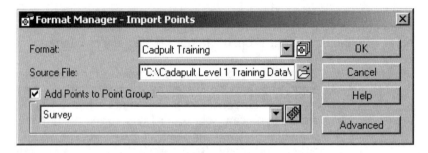

    2.   Select the **"Cadapult Training"** format.

    3.   Browse to the Source file:

C:\Cadapult Training Data\LDT Level 1\ Points\Site.txt

We also want to add the imported points to a point group.

    4.   Enable the **Add Points to Point Group** option, then click the green button (to the right of the drop down list) to create a new point group called **"Survey"**.

    5.   Click **<<OK>>** to close all dialog boxes. .

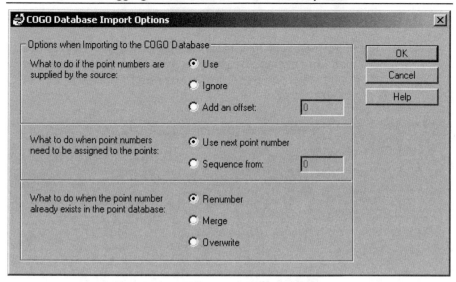

6. Click **<<OK>>** to use the default import options.

7. **Zoom Extents** to see the points you just imported.

4.3.2   Confirming The Description Keys Worked Properly

1. Open the **Layer Properties Manager** box to view the layers that were created by your description keys. If these new layers were not created then the description keys did not work properly.

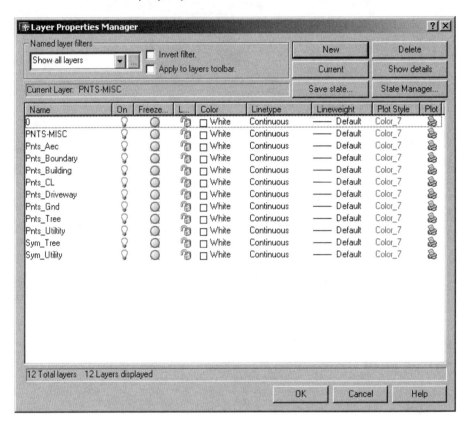

2. **Isolate** the layer **Pnts-Misc.** If you find any points on this layer these are the points that you did not have a description key properly defined for. The points are on this layer because it was current when you imported the points.

3. After reviewing the contents of the Pnts-Misc layer turn all of the other layers back on. You can do this with the **Layer Previous** command.

4.4   Changing The Point Display Properties

As you zoom in and look at your points you may notice that many of them overlap because of their size. There are two different ways to fix this problem. The first is to edit the display properties of the points that already exist in your drawing. This will fix the display problem that you currently have but it will not change the default settings for your drawing that control points that you will insert to the drawing in the future. So a second way to address this problem is to change the Marker and Text properties in the Point Settings.

4.4.1   Point Display Properties

This will edit the display of points you select that are currently in your drawing.

1. Select **Points** $\Rightarrow$ **Edit Points** $\Rightarrow$ **Display Properties**

2. At the command line type **A** when asked what points to modify to select all the points in the drawing.

```
Points to Modify (All/Numbers/Group/Selection/Dialog)?
<Dialog>: A
```

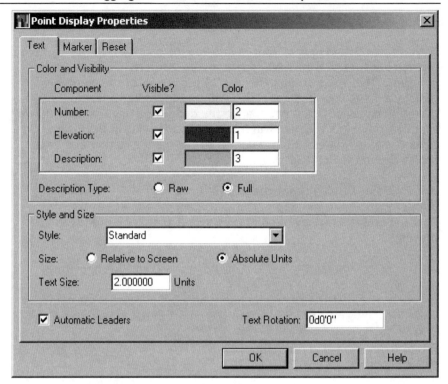

3. On the **Text** tab set the **Text Size** to **2**.

4. Turn on the option for **Automatic Leaders**.

5. You can also change the color and visibility of the point Number, Elevation, and Description if desired.

6. Select the **Marker** tab.

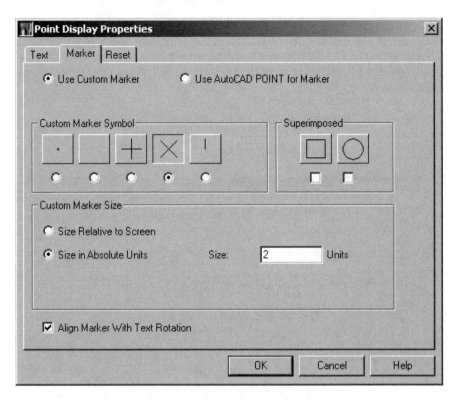

7. Set the marker Size to **2**.

8. Click **<<OK>>** to apply the changes to the points in the drawing.

4.4.2   Default Point Display Settings

You can set the default point display in the **Point Settings** command. This will control the display of the points as they are inserted into the drawing. These settings can also be saved into a Project Prototype so that all new drawings in a project that uses that prototype will have these settings by default.

1.   Select **Points** ⟹ **Point Settings**.

2.   Select the **Marker** tab.

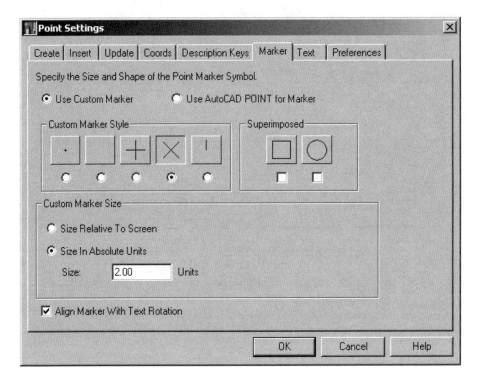

3.   Set the marker Size to **2**.

4. Select the **Text** tab.

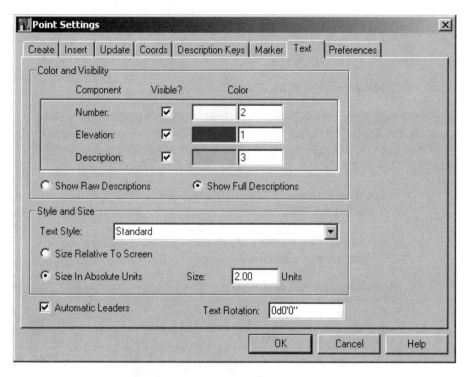

5. Set the **Text Size** to **2**.

6. Turn on the option for **Automatic Leaders**.

7. You can also change the color and visibility of the point Number, Elevation, and Description if desired.

8. Click **<<OK>>** to save these default settings for this drawing.

This sets the defaults for this drawing but it does not change the settings in the Project Prototype. To save changes like these to a Prototype you must use either the **Prototype Settings** or the **Edit Drawing Settings** command in the Projects pull-down.

4.5   Working With Point Groups

Point Groups are selection sets of points that we can save with the project. Once you define a Point Group, whether it is simple or complex, you can then select points by that group while using any point related command.

4.5.1   Managing Point Groups

Many common point commands are available by right-clicking on a point group in the Point Group Manager. These are the same commands that you can find in the Points pull-down, the Point Group Manager is just another way to access them.

1.   Select **Points** ⇒ **Point Management** ⇒ **Point Group Manager**.

2.   **Right-click** on the **Survey** point group and select **Lock Points in Point Group**.

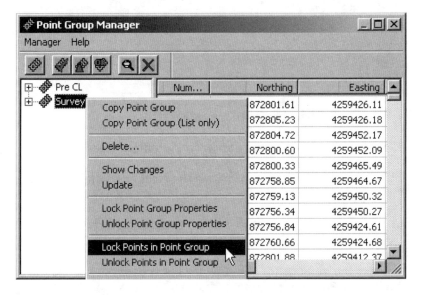

This locks all the points that are in the selected group so that they cannot be edited in the database. This is the same as using the **Points** ⇒ **Lock/Unlock Points** ⇒ **Lock Points** command from the **Points** pull-down.

3. Now **right-click** on the **Survey** point group and select ⇒ **Lock Point Group Properties.**

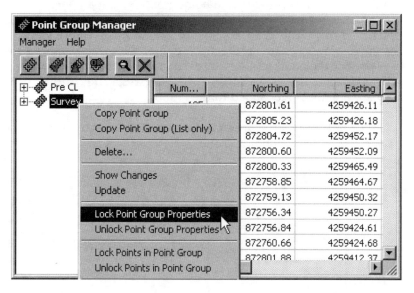

This locks the Properties of the point group so that points cannot be added or removed from the group. This is a good precaution to take with our survey group because we want to ensure that it contains all the points that came from the surveyor and no more. Once a point group's Properties are Locked the icon changes to include a padlock as shown below.

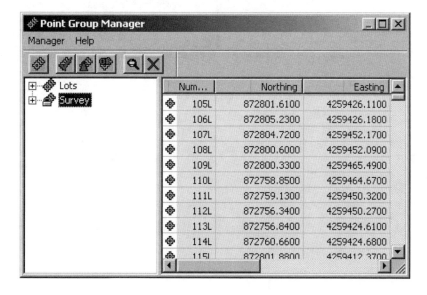

4.5.2   Creating a Point Group for Property Corners

In this exercise you will create a point group for the property corner points. This group will search for all the points in our project that use a certain description key. This is an example of a point group where you could save the group properties to a project prototype. This way when you start a new project the point group is automatically created. Then when points are imported to the project that fit into the properties of the point group they are automatically added to the group.

1. Click the **green button** on the left of the toolbar to **create a new point group.**

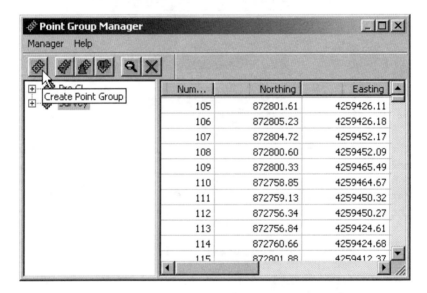

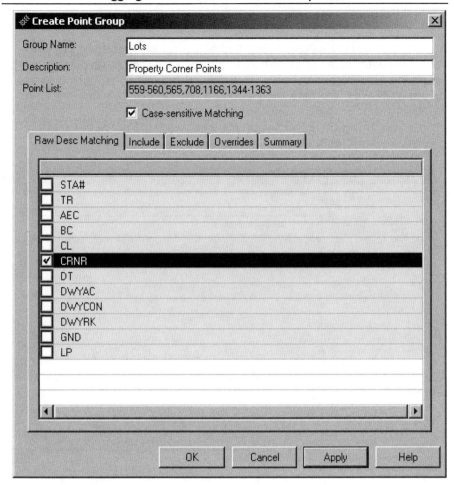

2. Enter a Group Name **"Lots"**.

3. Enter a Description of **Property Corner Points**.

4. Select the **"Raw Desc Matching"** Tab.

5. Select the Raw Description **"CRNR"**.

6. Click the **<<Apply>>** Button.

Now you will see the point numbers that match your filter in the Point List box.

7. Now click the **<<OK>>** button to save the group.

4.5.3   Create a Point Group for Center Line Points

You will now create a Point Group for the center line points by searching for a Raw Description of CL. You could also create the same point group by searching for a Description Key. However, this exercise will demonstrate an alternate way to create the group.

> 1.   In the **Point Group Manager** click the green button to create a new point group.

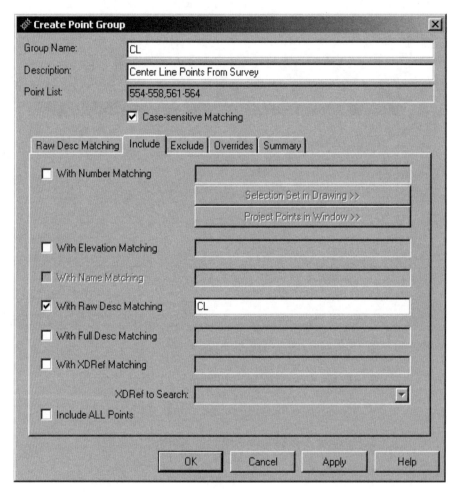

> 2.   Name the Point Group **"CL"**.
>
> 3.   Enter a Description of **Center Line Points From Survey**.

4. On the **"Include"** Tab setup a filter for the Raw Description **"CL"**.

5. Click **<<OK>>** to save the group definition.

4.5.4 Adding Points to the Drawing by Group

1. Select **Points** ⇒ **Remove From Drawing**.

2. Answer "**Yes**" to the command line question "Also remove Description Key symbols?".

3. Type "**A**" to remove all points.

If the points are locked, as they should be after the previous exercise, you will see the following dialog box.

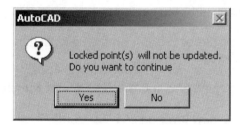

4. Click **Yes** to remove the points from the drawing.

5. Now Select **Points** ⇒ **Insert Points to Drawing**.

6. Type "**G**" to select Points by Group.

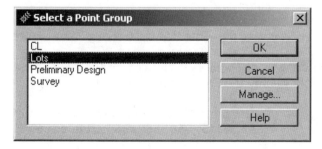

7. Select the Group "**Lots**" from the list.

8. Click **<<OK>>** to insert the points.

4.6   Drawing Linework Using the Land Desktop Lines/Curves Commands

The commands in the Lines/Curves pull-down draw 2D lines and arcs that are exactly the same objects as you create by using the AutoCAD line and arc commands. The difference is that the commands in the Lines/Curves pull-down are aware of the Land Desktop project so they can draw lines or arcs to the locations of the project points even if they are not currently in the drawing.

While using the Lines/Curves ⇒ Line command, as well as many other commands in Land Desktop where you are asked to specify a location, you can use the following command line toggles to specify the location by point number or a number of other methods shown below.

(.D dist/angle) (.R relative) (.I intersection) (.F fraction)
(.M midof) (.H hold fraction) (.GC GeomCal)
(.P point number) (.N northing and easting) (.G graphic point block)

Remember these are toggles so .P turns on the Point Number option and typing .P again turns the option off.

4.6.1   Using the Line command and the ".P" (Point Number) Option

1. Create a new layer named "**LOT LINES**", set it **Current** and color **Yellow**.

2. Select **Lines/Curves ⇒ Line**.

At the Command Line you will see the prompt "Starting Point".

3. Enter ".P".

This will allow you to draw the line by entering Point Numbers.

When entering point numbers in this command you must enter each point number followed by the **ENTER** key.

4. Enter points **1345, 1344, 1166, 1350, 1349, 1351, 1352, 1353, 1345, 559, 560, 565, 708, 1166.**

5. **ENTER** to end the line.

6. **ENTER** a second time to end the command.

4.6.2   Using the Line command and the ".G" (Graphical) Option

  1.   Select **Lines/Curves** ⇒ **Line**.

  2.   At the command line enter ".G".

Now you can graphically select the Point Objects from the screen. You can select a point by picking any part of the point object, not just the marker. This command does not give you a rubberband line from point to point like the normal AutoCAD Line command does. So be careful about picking the same point more than once. If you do you will draw a line from that point to the same point and create a zero length line.

  3.   Select Points **708, 1346, 1347, 1348, 1349.**

  4.   **ENTER** to end the line.

  5.   **ENTER** a second time to end the command.

4.6.3   Drawing Lines by a Range of Point Numbers

1. Select **Lines/Curves ⇒ By Point # Range.**

Now you can enter point number ranges using a dash for a range and a comma for gaps in the range.

2. At the command line enter: **1360-1363,1360.**

3. **ENTER** to draw the line.

4. Now enter: **1347,1354-1359,1355.**

5. **ENTER** to draw the line.

6. **ENTER** a second time to end the command.

7. Now use the **<<Lines/Curves>> <<Line>>** command to fill in the gaps in our linework.

8. Enter .G to turn off the Graphical prompt before exiting the Line command.

When finished the parcels should look like the graphic below.

4.7   Working With Parcels

Areas can be defined in Land Desktop and stored in a project database as Parcels. These areas may by parcels or tax lots by definition or they may be any other area that is important to your design such as areas defining the phases of construction. By defining an area into the parcel database it allows you to import the saved geometry into other drawings as well as create reports for that defined area such as area, inverse, and map check reports. Parcels are also used when calculating site volumes. If you want to calculate a volume for a specific, irregularly shaped area then you need to define that area as a parcel and calculate a parcel volume.

4.7.1 Parcel Settings

1. Select **Parcels** ⇒ **Parcel Settings**.

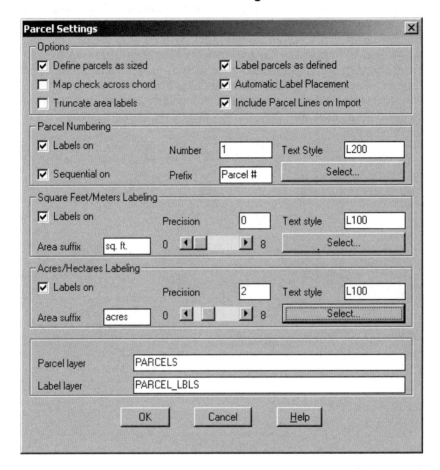

2. Set the Parcel Numbering Text Style to **L200** and the Parcel Number Prefix to "**Parcel #**".

3. Set the Square Feet Labeling Text Style to **L100**.

4. Set the Acres Labeling Text Style to **L100** and the precision to 2.

5. Click **<<OK>>** to save the changes to the Parcel Settings.

4.7.2 Defining A Parcel From Existing Geometry

Parcels can be defined by selecting a closed polyline, a group of lines and arcs, or a set of points. With any of these options the geometry you select must close or the parcel command will close it for you, and you may not like the results.

1. Isolate the "Lot Lines" layer.

2. Select **Parcels** ⇒ **Define from Lines/Curves**.

We will start by defining the parcel in the northwest corner of our site.

3. Pick the horizontal line in the northwest corner of the parcel near the west end. This will highlight the line and place a red X at the west end of the line. This is the Point Of Beginning in our parcel definition.

4. Now select the remaining lines that make up the parcel.

5. When you are finished selecting the lines **Enter** to define the parcel. If the lines you have selected do not close you will be asked if you would like to force closure.

6. After the first parcel has been defined the command restarts and you are ready to define your next parcel. Continue and define several more parcels.

4.7.3 Managing Parcels And Creating Reports

The Parcel Manager allows you to create reports, import, rename, and delete parcels from the project database.

1. Select **Parcels** ⇒ **Parcel Manager.**

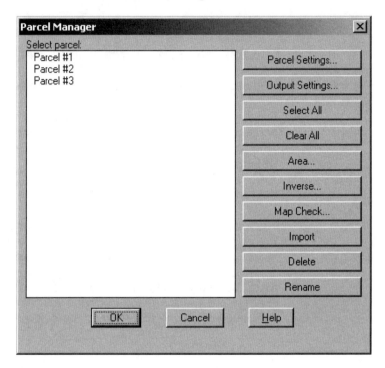

2. Select a Parcel or parcels and then select the desired option.

4.7.4 Parcel Sizing With The Slide Bearing Command

Land Desktop also includes several Parcel Sizing commands that can help you layout parcels and define them an exact size. This helps to take the guesswork out of parcel layout. If you want a parcel that is exactly 25,000 square feet, then you can create one automatically, rather than through trial and error with the area command.

When using the Slide Bearing command you will trace parcel lines that have already been created in your drawing that define the fixed sides of the parcel you are sizing. Then you will define the two directions that the closing parcel line will slide on. Finally you will be asked for the direction of the closing parcel line and the desired area of the parcel. Once you have supplied Land Desktop with all this information it will define the parcel and draw the new parcel line.

The Slide Bearing command, like most of the parcel commands, is run completely through the command line rather than dialog boxes. It also has no internal option to undo or back up a step. So if you make a mistake during the command your only option is to cancel the command and start over.

In the following exercise we will use the Slide Bearing command to define a parcel that has an area of 25000 square feet with a new parcel line that is perpendicular to the parcel line on the south side of the parcel.

1. Select **Parcels ⇒ Slide Bearing**.

2. Use the **End Point** object snap to pick the three points shown in the graphic below marked with a donut starting with the northwest point (the donuts have been added to the graphics in this exercise for illustration purposes, they will not appear in your drawing unless you decide to add them). This step has traced fixed lines into the new parcel and will determine the minimum size of the parcel that we will be shown later.

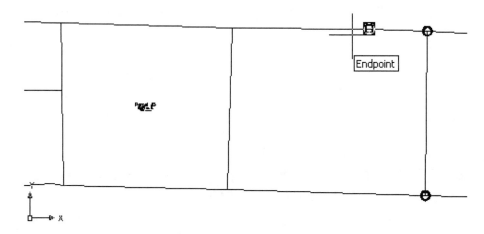

3. After picking the last point, hit **Enter** to move to the next step in the command.

4. Now you are given a rubber band line from the first parcel point and asked to pick the **First direction**: use a nearest object snap to pick a point on the line that is connected to the first parcel point you selected as shown in the graphic below. This defines the first of two direction lines that our new parcel line will slide on to find the desired area.

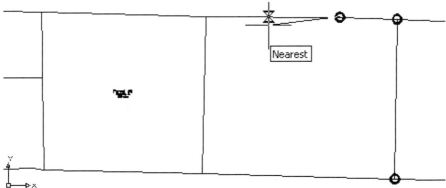

5. Next you are taken to the last parcel point that you picked during the tracing portion of the exercise, given a rubber band line from this point and asked to pick the **Second direction**: use a nearest object snap to pick a point on the line that is connected to the last parcel point you selected as shown in the graphic below. This defines the second of two direction lines that our new parcel line will slide on to find the desired area.

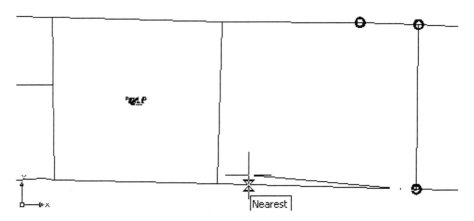

6. You are now asked to define the direction of the new parcel line that will close our parcel. You can enter the direction of this line by typing in the Quadrant, bearing, and distance (similar to the By Direction command in the Lines/Curves pull-down that we used in an earlier exercise); by typing an A and then entering an Azimuth; or by entering PO and picking two locations in the drawing with an object snap to graphically determine the direction. We will use this last option so enter **PO** at the command line and press **Enter**.

7. This will change your command prompt to >**First point**: use a **nearest** object snap to pick a point on the line that we defined as our first direction line as shown below.

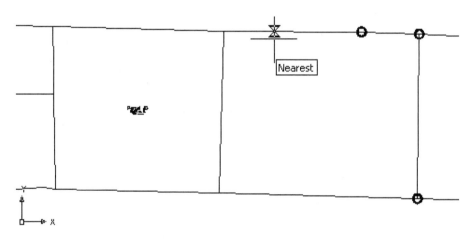

8. Now you are asked for the **>Second point**: to define the direction of the closing parcel line. Use a **perpendicular** object snap and pick a point on the second direction line as shown below.

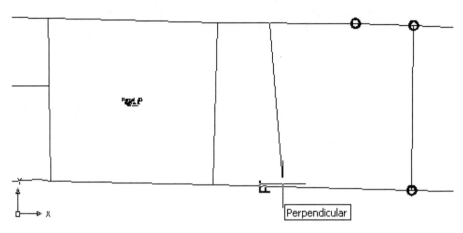

9. Once the direction of the closing parcel line is defined a minimum parcel area is displayed at the command line and you are prompted to enter the **Desired area, in square units**: Enter **25000** and press **Enter**.

10. The parcel is now sized, defined into the Parcel Database, a new parcel line is drawn in the drawing on the current layer, and the parcel is labeled according to the Parcel Settings. The command then starts again so if you want to size another parcel you can. **Enter** to end the command.

4.8   Labeling Linework

Land Desktop uses a style-based method of labeling lines, curves, spirals, and points. The label style controls all the properties of the label including what geometric data is labeled as well as the layer and text style of the label text. When this method is used properly it can help improve compliance with standards and ease of use.

Many of the commands shown in the exercises below are selected from a right-click popup menu. This is only one way to access these commands. If you prefer, you can find all of the label commands in the Labels pull-down menu.

In the following exercises you will be labeling lines. However, curves and spirals can be labeled the exact same way. You just need to be sure that your label styles are defined properly for each type of object so that the labels are created in a consistent way.

Tables can also be created to display Line, Curve, or Spiral labels. If you decide to create a table instead of labeling on the geometry then the geometry is tagged; for example the first line will be tagged with L1, the second with L2, and so on. This is a good option to use if your geometry is short enough that standard labels do not fit on the object.

4.8.1   Creating A Line Label Style

1. Select **Labels** ⇒ **Show Dialog Bar.**

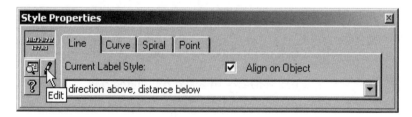

2. Click the **Pencil** button to edit the current style.

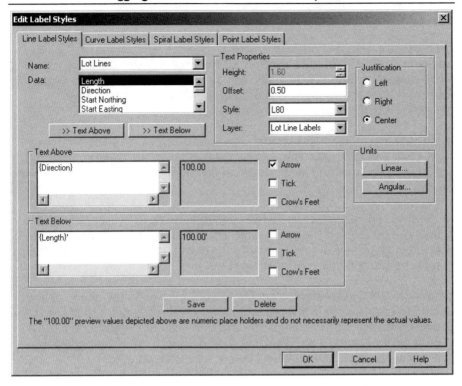

3. Enter a new Label Style Name of **"Lot Lines"**.

4. Set the Text Style to **"L80"**.

5. Type in **"Lot Line Labels"** for the Layer Name.

6. Click the **<<Save>>** button to save the new style.

7. Click **<<OK>>** to leave the dialog box.

4.8.2 Labeling Lines

1. Set **"Lot Lines"** as the current Line Label Style.

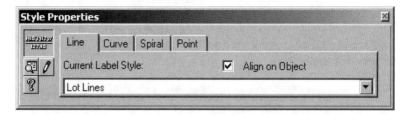

2. Now pick several lines that you would like to label and **right-click**.

3. Select ⇒ **Add Dynamic Label** from the right-click menu.

Since these labels are dynamic they automatically update if you edit the line or if you edit the label style.

If you do not want your labels to be dynamic you can create Static Labels using the same label style by selecting Add Static Label when right-clicking on the object.

4. If you want to erase the label select **Labels** ⇒ **Delete Labels**.

5. If you want to change the direction that a line is labeled select **Label** ⇒ **Flip Direction**.

6. You can use **Labels** ⇒ **Swap Label Text** to switch the text to the other side of the line.

7. You can make a Dynamic Label Static by selecting **Labels** ⇒ **Disassociate Labels**.

4.8.3 Creating A Line Tag Style

1. Click the button on the upper left corner of the **Style Properties** dialog box to toggle from Label to Tag mode.

This is not necessary to create the tag labels. It is only required to set the current tag label style or to edit the tag label style through the Style Properties dialog box.

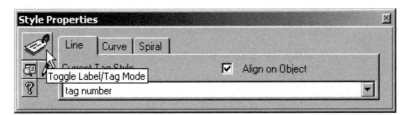

2. Select the **Pencil** button to edit the Tag Style.

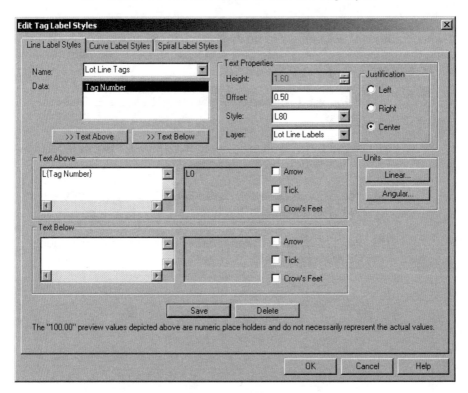

3. Enter a new Label Style Name of **"Lot Line Tags"**.

4. Set the Text Style to **"L80"**.

5. Type in **"Lot Line Labels"** for the Layer Name.

This layer will already exist and can be selected from the list if you created labels in the previous exercise.

6. Click **<<Save>>** to save the new style.

7. THEN, click **<<OK>>** to leave the dialog box.

4.8.4   Tagging Lines

1. Set **Lot Line Tags** as the current Tag Style

2. Now pick several lines that you would like to tag and **right-click.**

3. Select ⇒ **Add Tag Label** from the right-click menu.

4.8.5 Creating A Line Table

1. Select **Labels** ⇒ **Add Tables** ⇒ **Line Table**.

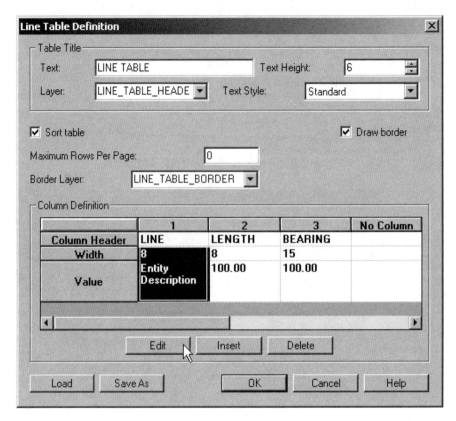

2. The **Line Table Definition** dialog box will be displayed.
Here you can control the display and the format of table.
Select column 1 and then click the **<<Edit>>** button to edit
the format of the first column in the table.

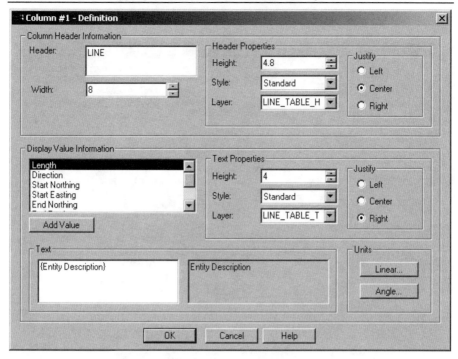

3. Here you can make any desired changes to the format of the selected column. Click **<<OK>>** when you are finished to save your changes and return to the **Line Table Definition** dialog box.

4. Edit any other columns as desired.

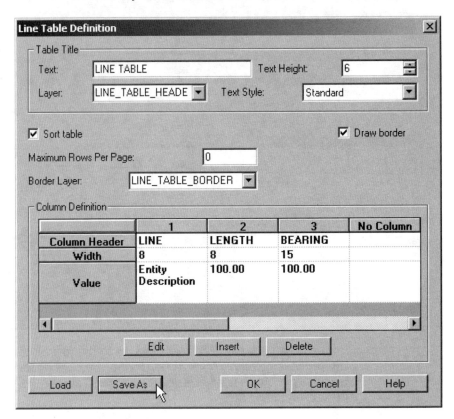

As you can see there are a lot of details that can be configured as part of your table definition and the setup can be time consuming. So once you take the time to configure the table definition exactly the way you like, it is a good idea to click the <<SaveAs>> button to save the table definition into the Label Library for future use.

5. Click <<OK>> to create the table.

6. At the command line you are asked to **Select Table Insertion Point** : Pick a point on the screen where you would like to create the table.

LINE TABLE		
LINE	LENGTH	BEARING
L1	190.10	N88°08'13"W
L2	227.32	N00°15'07"E
L3	192.50	S89°54'17"E

There is not a dynamic option for Tables like there is for regular labels. So if you make any changes to the tagged geometry you must update the table.

7. If you want to update the table with any changes that may have occurred to the geometry of the tagged lines Select **Labels ⇒ Edit Tables ⇒ Re-Draw Table.**

8. Then pick any line or text in the table and the entire table will be redrawn with the current data.

9. Tables can be erased with the AutoCAD Erase command or by selecting **Labels ⇒ Edit Tables ⇒ Delete Table.** With the Delete Table command you will be prompted to pick any line or text in the table and the entire table will be erased.

4.9   Chapter Summary

In this chapter you saw how to use point groups and description keys to help manage your point data. Because of their importance they both should be saved into a project prototype. This will add consistency to your projects as well as save you the time of setting them up again. We also looked at how to use the COGO, labeling, and parcel commands to speed up drafting, as well as some of the design components of your project.

Chapter 5

Build And Define A Survey Quality Existing Ground Model

In this chapter we will use points and breaklines from our survey data to create a survey quality existing ground surface. We will look at ways to leverage the use of Point Groups to efficiently build and edit our surface. We will also explore various ways of editing and analyzing surfaces including the use of our preliminary surface to add extra data beyond the limits of our survey.

- **Building A Surface From Survey Data**

- **Editing The Surface**

- **Surface Analysis**

- **Contours**

Dataset:

To start this chapter we will continue working in the drawing named **Survey.dwg.** You can continue with the drawing and project data that you currently have from the end of the previous chapter or you can install the dataset named **"Chapter 5 Cadapult Level 1 Training.zip"** from the CD that came with the book. Extract the dataset to the folder **C:\Land Projects 2004** or whatever folder you have been using as your project path. Extracting the project from the dataset provided will ensure that you have the project and drawings set up correctly for the exercises in the following chapter and overwrite any mistakes that you may have made in previous exercises.

5.1   Building A Surface From Survey Data

Any time you build a surface the most important step is to understand what data you have available to work with. In this chapter we will work with Points that we will manage with a point group and Breaklines that we will create based on some of those same survey points.

5.1.1   Surface Display Settings

We set these same Surface Display Settings in a previous chapter for the Preliminary Design drawing. However, since we did not save them to the project prototype they were only saved in that drawing. We will need to set them again in the following exercise. This illustrates the importance of setting up and using a project prototype.

1. Open the drawing **Survey.dwg** from the project **Cadapult Level 1 Training** if it is not already open.

2. Select **Projects** ⇒ **Edit Drawing Settings**.

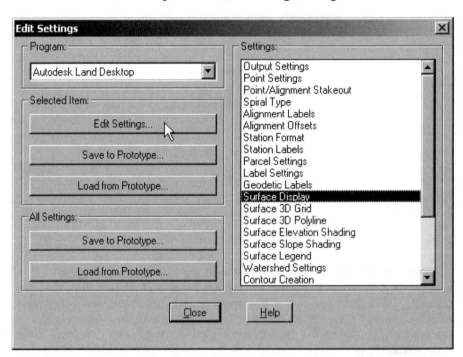

3. Select **"Surface Display"** from the list and click **<<Edit Settings>>**.

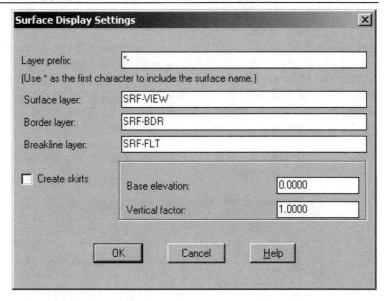

4. Enter an **"*—"** as the Layer Prefix.

This will use our surface name as a prefix for all of our surface layers.

5. Click **<<OK>>** and **<<Close>>** to exit all dialog boxes.

5.1.2   Creating A Point Group To Be Used As Surface Data

Before we create the surface we need to create a point group that we will use to select only the points that we want to use for our surface data. Points that should not be included in the surface should not be included in the point group. Points for utility potholes or points that are part of the project for horizontal control and do not have accurate elevations are examples of points that should not be included in this group.

1.   Select **Points** ⇒ **Point Management** ⇒ **Point Group Manager**

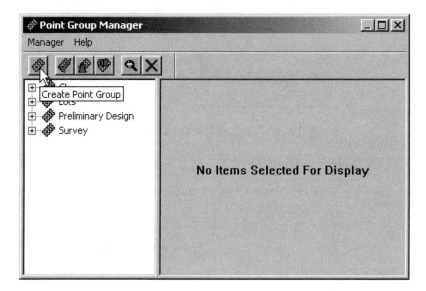

2.   Select the **green button** on the left to create a **new point group**.

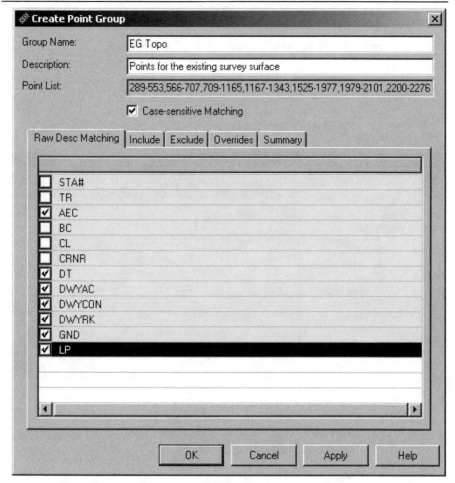

3. Name the group **"EG Topo"**.

4. On the **"Raw Desc Matching "** tab select the description keys **"AEC, DT, DWYAC, DWYCON, DWYRK, GND,** and **LP"**.

5. Click the **<<Apply>>** Button.

Then you will see the point numbers that matched your filter in the Point List box.

6. Click **<<OK>>** to save the new point group.

7. Close the point group manager.

5.1.3 Creating The Survey Surface

1. Select **Terrain** ⇒ **Terrain Model Explorer**.

2. **Right-click** on the **Terrain** folder and select ⇒ **Create A New Surface**.

3. **Right-click** on **"Surface 1"** and rename it to **"Survey"**.

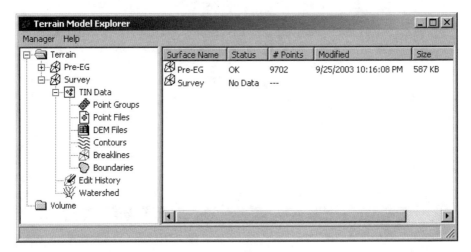

5.1.4 Adding Point Group Data To A Surface

1. **Right-click** on **Point Groups** under the **Survey** surface and select ⇒ **Add Point Group**.

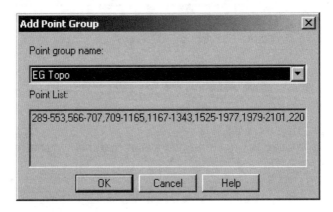

2. Add the Point Group **"EG Topo"**.

5.1.5   Defining Breaklines By Point Number

Breaklines are a very important part of defining an accurate surface because they force surface triangulation along the breakline and prevent triangulation across it. Typical places that you should consider using breaklines are tops and toes of slopes, curbs, edges of pavement, crowns of roads, and retaining walls. Basically any place that you have a change in grade you should think about using a breakline.

This command allows you to type in a list or range of point numbers that Land Desktop will use to define a breakline. This is similar to the By Point # Range command in the Lines/Curves pull-down that we used in a previous chapter. The points do not even need to reside in the drawing because the command looks at the point database in the project for the needed information rather than looking in the drawing.

After entering the point numbers you will be asked for a description of the breakline. This is an optional step and can be skipped with an ENTER. The description can be useful when editing or looking for a specific breakline but it is not required and often skipped. If you do not give the breakline a description it will have a default description of "Unclassified".

Finally a 3D polyline will be drawn on the breakline layer. This layer name is controlled by the Surface Display Settings and can be changed by selecting **Projects** ⇒ **Edit Drawing Settings** then selecting **Surface Display** from the list.

1. **Right-click** on **Breaklines** and select ⇒ **Define by Point Number.**

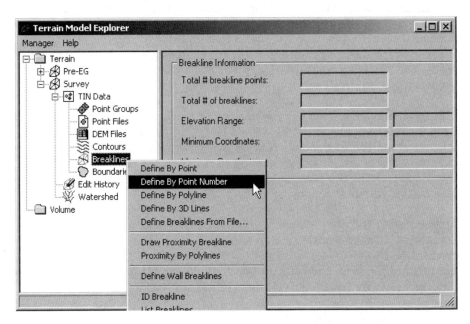

2. At the command line enter the number ranges and descriptions in the following table to define the breaklines.

Point Numbers	Descriptions
289-410	Road 1
412-553	Road 2
546,566-707	Road 3
551,709-748	Road 4
749-821,992,991,990,749	Road 5
1022,993-1021,822-916	Road 6
917-970,917	Road 7
1546-1668	Road 8
1716-1807	Road 9

3. **ENTER** to end the command.

4. Close the terrain Model Explorer when you are finished.

5.1.6 Defining Breaklines By Point Selection

This command will allow you to define breaklines by selecting project points one at a time from the drawing. As you pick each point a temporary 2D line is drawing on the current layer. Once you finish picking the points the temporary line is erased and replaced by a 3D polyline on the breakline layer.

1. Currently the only points that are in the drawing are the Property Corner points. Select **Points ⇒ Insert Points To Drawing.**

2. At the command line type **G** to select the points by Group.

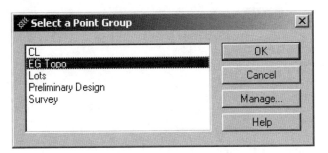

3. Select the group **EG Topo** to insert the points that we are working with for our surface.

4. Isolate the layers **PNTS_DRIVEWAY and Survey-SRF-FLT.**

5. Set layer **Pnts_Driveway** current.

6. The point that we will be starting our breakline from is Point number 1139. To locate that point select **Points ⇒ Point Utilities ⇒ Zoom To Point.**

7. At the command line you will be prompted for a point number. Enter point 1139. **Point to zoom to: 1139**

8. Next at the command line you will be prompted for a **Zoom height <1162.08>: 75** This is a distance displayed vertically on your screen. So the smaller the number the closer you are zoomed in to the selected point.

9. Select **Terrain** ⇒ **Terrain Model Explorer.**

10. **Right-click** on **Breaklines** under the **Survey** surface in the Terrain Model Explorer and select **<<Define By Point>>**

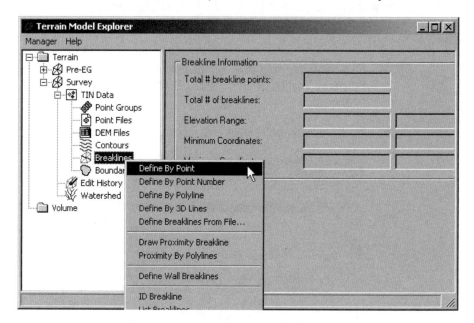

Now you will be taken to the command line and prompted to **Select first point:**

11. **Pick** point **1139** from the screen.

12. Then pick points **1141, 1140, 1146**, and **1765**.

This selection method is similar to using the Line command from the Lines/Curves pull-down with the Graphical (.G) option that we covered in Chapter 4 section 4.6.2.

13. **ENTER** when you are finished selecting points.

14. Enter a breakline description of **Road 10**.

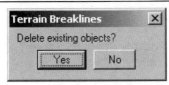

15. Click **<<Yes>>** to delete the existing objects. This will erase the temporary line that has been drawn during the command and replace it with a 3D brakline on the breakline layer.

After completing the first breakline the command restarts and asks you for the First Point of the next breakline.

16. Starting at point **1136** define a second breakline with a description of **Road 11** along the east side of the driveway using points **1136, 1138, 1137, 1144, 1145**, and **1767**.

17. **ENTER** to end the command and return to the Terrain Model Explorer.

5.1.7 Editing Breaklines

1. Select **Points** ⇒ **Point Utilities** ⇒ **Zoom To Point** and zoom to point **411** at a height of **50**.

2. **Right-click** on **Breaklines** under the **Survey** surface in the Terrain Model Explorer and click **<<Edit Breakline>>**.

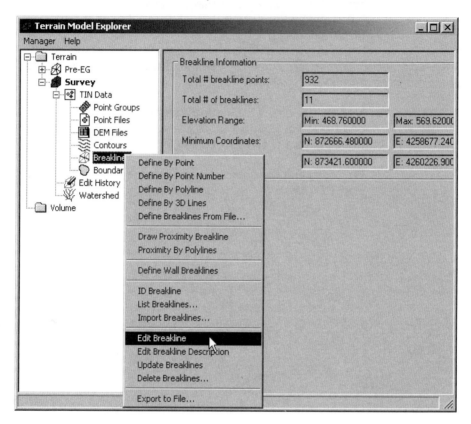

3. Select breakline "Road 1" between points 409 and 410.

The command line displays the following prompt:
```
Breakline #: 1  Description: road 1  Type: Standard
Current elevation =   469.04
Next/Previous/eXit/Move/Elevation/Insert/Delete <Next>:
```

This is similar to editing the vertexes of a polyline in the PEDIT command. However, in the Edit Breakline command the command always starts at the first point on the breakline whether it is visible on screen or not. So in our example the selected vertex is off the screen a long distance to the west.

4. **ENTER** to use the default command line option of **Next**. This moves the command to the second vertex of the breakline.

5. Continue selecting the **Next** option until you get to the last point on the breakline. It will display a current elevation of 560.88 at the command line. You would normally see a graphical X move from vertex to vertex as you use the next and previous options. But in our example the points are using a marker that is displayed with the X symbol so the point marker obscures the X from the edit breakline command.

6. At the command line enter **I** to insert a new vertex into the breakline.

7. Use a **Node** object snap to select point **411** when you are asked for a **New vertex location:**

This only selects a horizontal location. It does not read the elevation of the point from the point database.

8. In the next step you are asked to enter the elevation of the new breakline vertex.
    New elevation <560.88>: 561.62

9. Now type an X to exit the Edit Breakline Command.

10. **ENTER** to end the command and return to the Terrain Model Explorer.

5.1.8 Building the Surface

> 1. **Right-click** on the surface **Survey** in the Terrain Model Explorer and select ⇒ **Build**.

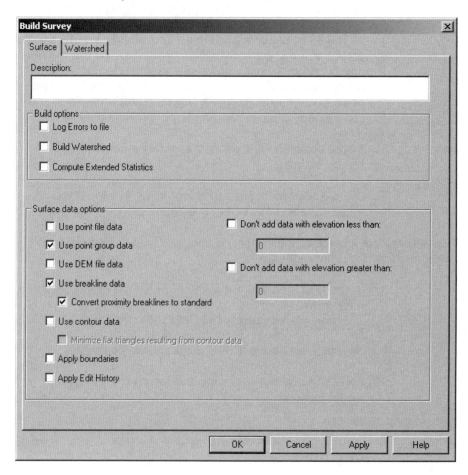

> 2. Build the surface from **Point Group Data** and **Breakline Data**.

If other options are selected it will not cause a problem as long as that data does not exist. Foe example if the option for DEM data was selected Land Desktop would look for any DEM data that had been added to our surface, not find any, and move on without causing any problem.

5.1.9 Viewing the Surface

1. **Right-click** on the surface **Survey** and select **Surface Display ⇒ Polyface Mesh.**

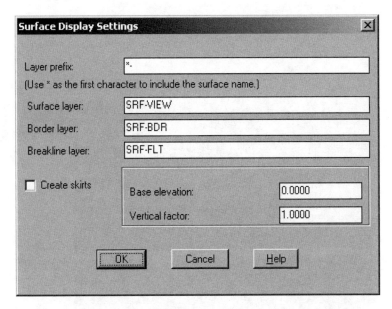

2. You will be asked to confirm the Surface Display Settings that we setup earlier in the chapter. Click **<<OK>>.**

3. At the command line answer **Yes** when asked to erase the old surface view layer.

The polyface mesh representing the surface is now displayed.

4. Close the Terrain Model Explorer.

5. **Zoom Extents.**

6. Select the Polyface Mesh and **Right-click**.

7. Use the **Object Viewer** to view the Polyface Mesh.

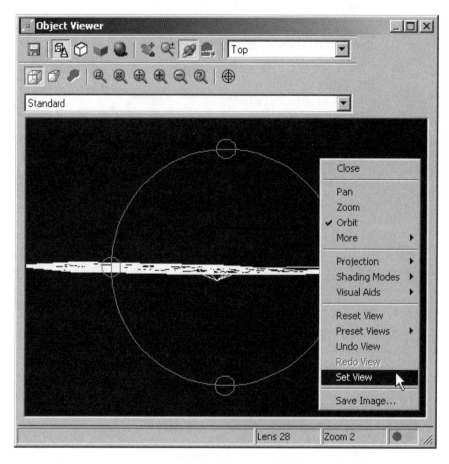

In the Object Viewer you will notice there is a spike in our surface.

8. When you have rotated the surface in the Object Viewer so the spike is exposed **Right-click** and select the **Set View** option.

9. Now **close** the **Object Viewer** and you will notice the AutoCAD view has changed to match the angle that was displayed in the Object Viewer.

10. **Explode** the Polyface Mesh.

11. Change the color of the 3D faces that make up the spike to **magenta**.

12. Change back to Plan View by typing **Plan** at the Command line and returning twice.

You should now be able to see the problem area in the surface.

13. Turn on your point layers and identify the problem point.

You will find that point number 1892 caused the spike.

14. Select point **1892**, **right-click**, and select **Edit Points**

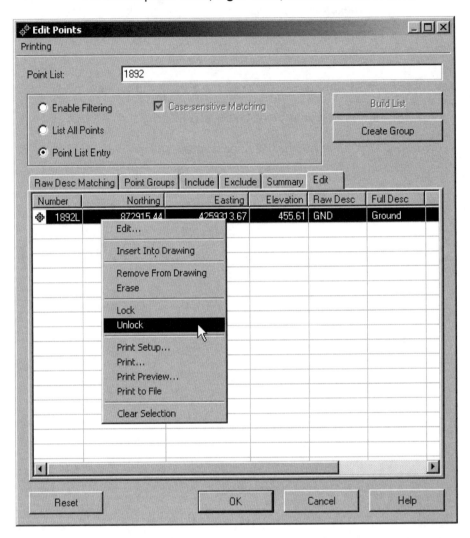

15. You will need to Unlock the point before you are allowed to edit it. Unlock the point by **Right-clicking** on it in the Edit Points dialog box and selecting ⇒ **Unlock.**

16. Change the elevation of point 1892 to **482.61.**

17. Lock the point after you are finished editing by **Right-clicking** on it in the Edit Points dialog box and selecting ⇒ **Lock.**

18. When you Lock the point you will be asked if you wish to save your changes, click **<<Yes>>**.

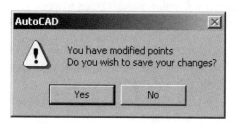

19. Click **<<OK>>** to close the Edit Points dialog box.

20. Open the Terrain Model Explorer by selecting **Terrain** ⇒ **Terrain Model Explorer.**

21. Now rebuild the surface by **Right-clicking** on the surface **Survey** and selecting **Build.**

22. You can now display a new polyface mesh showing the revised surface and view it in the object viewer as we did earlier in this exercise.

23. Close the Terrain Model Explorer.

5.2   Editing The Surface

A surface can be edited only if you have 3D lines in your drawing. The surface editing commands do not work with 3D faces. When editing a surface it is important to use the Edit Surface commands in the Terrain pull-down. These commands affect both the drawing and the surface database in the project while the standard AutoCAD commands only act on the drawing and do not edit the database.

5.2.1   Deleting Lines

1.   Select **Terrain** ⇒ **Edit Surface** ⇒ **Import 3D Lines.**

2.   **Erase** the surface view layer when prompted.

3.   Select **Terrain** ⇒ **Edit Surface** ⇒ **Delete Line.**

4.   Pick the long lines along the edge of our surface that have triangulated outside of our survey data.

Long skinny triangles are typically incorrect and need to be deleted. As you look at the TIN lines that connect each point ask yourself, "Should these points be associated with each other in the surface?" A line between two points will create a straight slope 3 dimensionally from one point to the other.

5.   When finished editing the surface select **Terrain** ⇒ **Save Current Surface.**

It is important after any Surface editing command is completed that you save the surface. If you do not save the surface you will be prompted to save the surface when you exit the program or open a different project. After editing the outer edge of the surface line by line you can see the value of creating a good surface boundary as we did when we created the preliminary existing ground surface. The boundary could have eliminated all of this surface editing.

6.   Once you are done editing the surface we can erase the 3D lines from the drawing. Select **Terrain** ⇒ **Terrain Layers** ⇒ **Surface Layer.**

7.   At the command line **ENTER** to select the default option of **Erase.**

This erases all of the objects on the layer **Survey-SRF-VIEW.**

5.2.2 Pasting Surfaces

Our final step is to combine the survey surface with our preliminary existing ground surface. We will use the preliminary existing ground surface as buffer data around our site-specific survey. We will do this with the Paste Surface command.

1. In the Terrain Model Explorer **right-click** on Surface **Pre-EG** and select ⇒ **Save As**.

2. Name the new surface **EG**.

3. **Close** the Terrain Model Explorer.

4. Select **Terrain** ⇒ **Edit Surface** ⇒ **Paste Surface**.

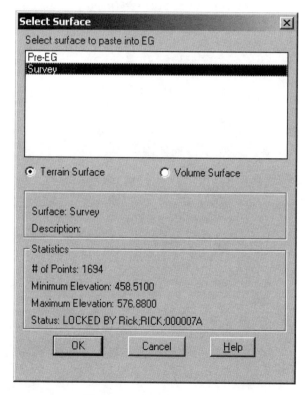

5. Select **Survey** as the surface to paste into **EG**.

6. Select **<<OK>>** to paste surface Survey into surface EG.

7. Select **Terrain** ⇒ **Save Current Surface**.

Build and Define a Survey Quality Existing Ground Model | 163

5.3   Surface Analysis

Once a surface has been created you have several options for surface analysis. You have previously viewed 3D faces and 3D lines, which is a basic form of surface analysis. Two other common types of surface analysis are Elevation Banding and Slope Analysis. Our exercises will look at the Slope Analysis commands; however, the Elevation Banding commands work in a very similar way.

5.3.1   Creating a 3D view of a Slope Analysis

Using 3D faces for the slope analysis is a good option if you plan on displaying the surface in a 3D or isometric view. It does not display very well in plan view. We will cover that in the next exercise.

1.   Select **Terrain** ⇒ **Surface Display** ⇒ **3D Faces.**

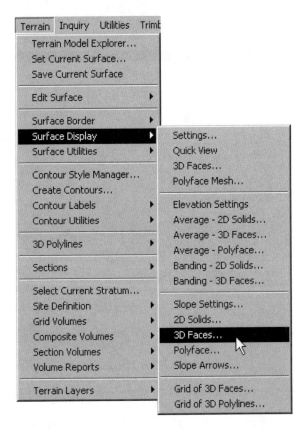

The 3D Faces option is available several places in this menu so be sure to select the option in the Slope section. This command is also available in the Terrain Model Explorer.

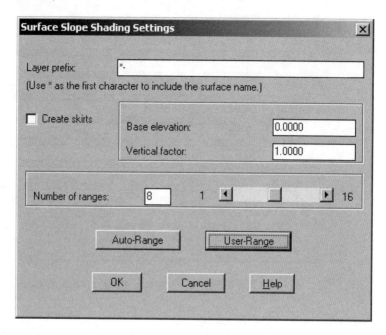

2. Enter **"8"** for the Number of Ranges.

3. Click **<<User-Range>>**.

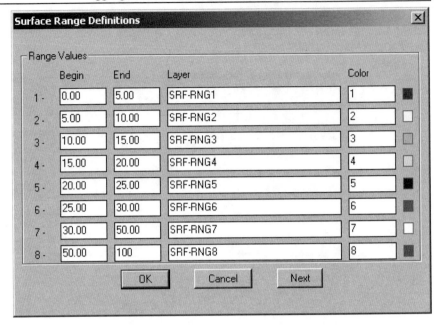

4. Set the Slope Ranges and their colors as desired.

5. Click **<<OK>>**.

6. Click **<<OK>>** in the Surface Slope Shading Settings dialog box to begin the Analysis.

7. Answer **"Yes"** at the command line to **Erase any Range layers** in the drawing.

You are then shown a dialog box that displays a summary of the slope analysis for the surface.

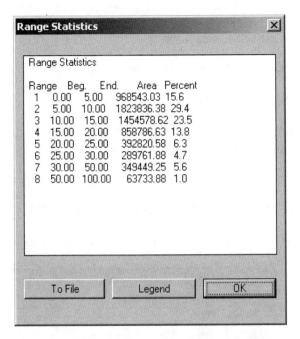

You have an option to save this information to a file and create a legend in the drawing.

8. Click **<<OK>>** to bring the 3D faces into the drawing.

9. **Zoom Extents.**

10. Isolate the following Range layers:

> **EG-SRF-RNG1**
> **EG-SRF-RNG2**
> **EG-SRF-RNG3**
> **EG-SRF-RNG4**
> **EG-SRF-RNG5**
> **EG-SRF-RNG6**
> **EG-SRF-RNG7**
> **EG-SRF-RNG8**

11. Select the 3D faces with a crossing window.

12. **Right-click** and select the ⇒ **Object Viewer** to view the slope analysis of the surface.

13. Since the objects are 3D faces they can be shaded in the object viewer for better display. **Right-click** in the object viewer and select ⇒ **Shading Modes** ⇒ **Flat Shaded** to shade the objects.

14. Close the Object Viewer.

5.3.2   Creating a Plan view of a Slope Analysis

For display in plan view doing the slope analysis with 2D solids is a better option. This gives you colored triangles that look similar to a solid hatch.

1.   Create a new **layer** named **Legend** and set it current.

2.   Select **Terrain** ⇒ **Surface Display** ⇒ **2D Solids.**

This command is also available in the Terrain Model Explorer

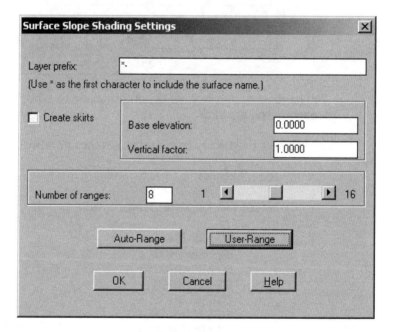

3.   Enter **"8"** for the Number of Ranges.

4.   Click **<<User-Range>>**.

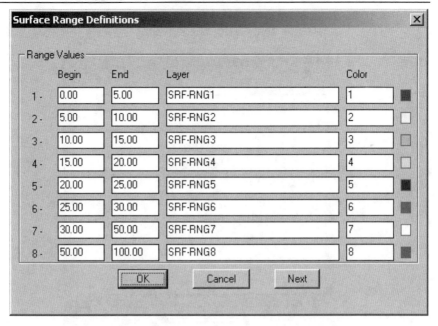

5. Set the Slope Ranges and their colors as desired.

6. Click <<**OK**>>.

7. Click <<**OK**>> in the Surface Slope Shading Settings dialog box to begin the Analysis.

8. Answer "**Yes**" at the command line to **Erase any Range layers** in the drawing. This will erase the 3D faces created in the previous exercise.

9. You are then shown a dialog box that displays a summary of the slope analysis for the surface.

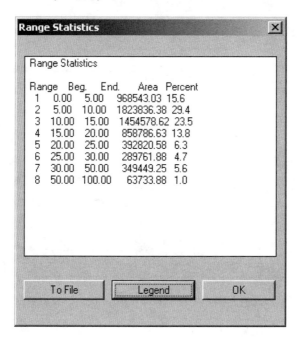

You have an option to save this information to a file and create a legend in the drawing.

10. Select the **<<Legend>>** button.

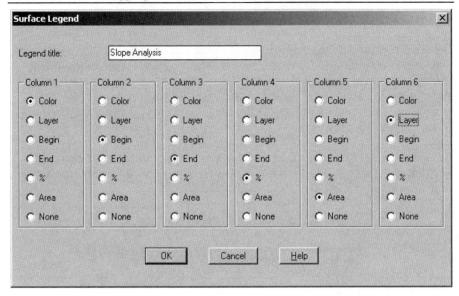

11. Enter **"Slope Analysis"** as the Title for the Legend.

12. Pick the information that you would like to display in each column of your legend.

13. Click **<<OK>>** to bring the 2D Solids into the drawing.

14. You will then be asked to **pick the insertion point**, or the upper left corner, of the legend. Pick a point somewhere to the right of your site.

The legend will be drawn on the current layer.

15. If you want to erase the Range layers select **Terrain** ⇒ **Terrain Layers** ⇒ **Range Layers.**

16. At the command line, **ENTER** to accept the default of **Erase.**

17. **Erase** the Legend with the AutoCAD Erase command.

5.4   Contours

Creating contours is just another form of surface display. In this section we will create contours, label them, and create a contour style that we will use to control the display of the contours. We will also save that contour style to the contour style library for future use as a standard.

5.4.1   Creating Contours

Before you start it is important to have the correct surface set current. For this exercise we will create contours from the Survey surface on a two-foot interval.

1.   Select **Terrain** ⇒ **Set Current Surface**.

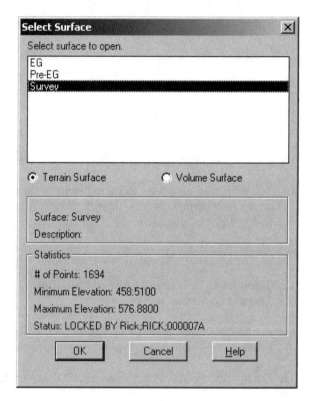

2.   Pick the **"Survey"** surface.

3.   Select <<**OK**>> to set the surface Survey current.

4. Select **Terrain** ⇒ **Create Contours**.

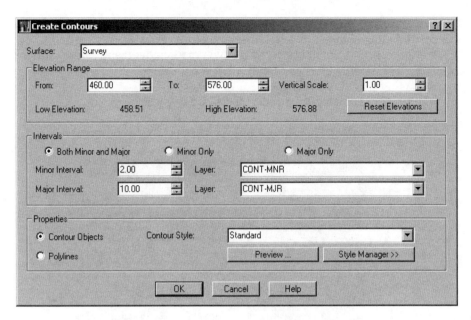

5. Be sure the current surface is set to **"Survey"**

6. Set a Minor Contour Interval of **2** and a Major interval of **10**

7. Create the contours as **Contour Objects**

8. Click **<<OK>>** to create the contours.

9. At the command line accept the default to **Erase** any old contours.

10. The contours were created on the layers SURVEY-CONT-MJR and SURVEY-CONT-MNR. Color the major contour layer **Green** and the minor contour layer **Yellow**.

5.4.2 Labeling Contours

It is important to isolate the contour layers before labeling because the contour labeling commands label lines and plotlines as well as contours. So if other geometry is available on screen you may inadvertently break other objects like property lines and label them with a zero.

1. Isolate the Contour layers and Layer 0.

2. Select **Terrain** ⇒ **Contour Labels** ⇒ **Group Interior.**

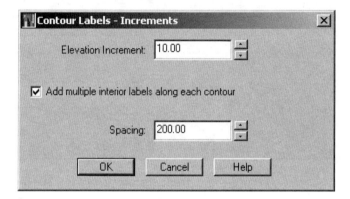

3. Set the Elevation Increment to **"10"**.

4. Enable the option to **"Add multiple interior labels along each contour"**.

5. Set the Spacing to **"200"**.

6. At the command line you are asked to pick a Start Point. Draw a line across the contours where you would like them labeled.

7. **ENTER** to end the labeling command.

You may not see any labels at this point. However, they are there. The reason that you cannot see them is that the contour style is not setup properly. We will fix this in the next exercise.

5.4.3 Contour Styles

Contour Styles control the appearance of the contour and the contour labels. In this exercise we will create a contour style to properly display the contours in the drawing. We will also save that contour style to the contour style library for use with other drawings in the future. Using a predefined contour style when you create contours lets you create contours that display consistently from drawing to drawing in an efficient way.

1. Select one of the contours and **Right-click**.

2. From the Right-click menu select **Contour Properties**.

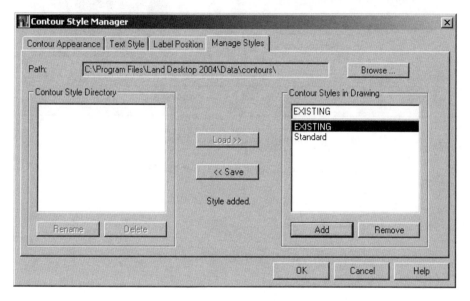

3. On the **Manage Styles** tab enter a **Style** name of **"Existing"**.

4. Click **<<Add>>**.

This will add a new contour style to this drawing. All of our changes will be saved in this style. Later we will export this style for future use in other drawings.

5. Select the **Contour Appearance** tab.

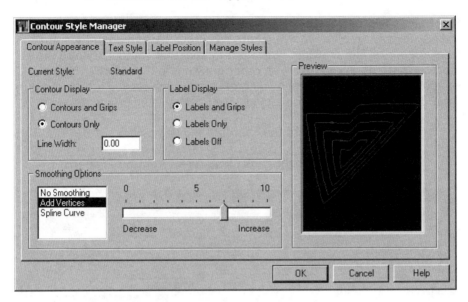

6. Pick the Smoothing Option of **"Add Vertices"**.

7. Increase the smoothing amount to **"7"**.

8. Select the **Text Style** tab.

The Text Style tab controls the display of the contour label text. Since the contour label is part of the contour line this is the only way that you can change the text style, height, or color.

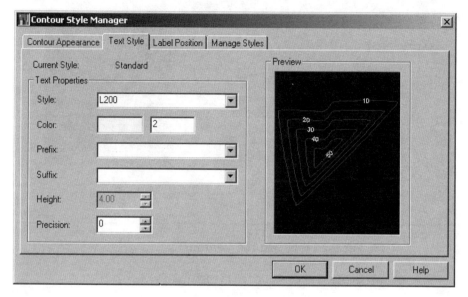

9. Set the Text Style to **"L200"**.

10. Set the color to **Yellow**.

11. Set the Precision to **"0"**.

12. Pick the **Label Position** tab.

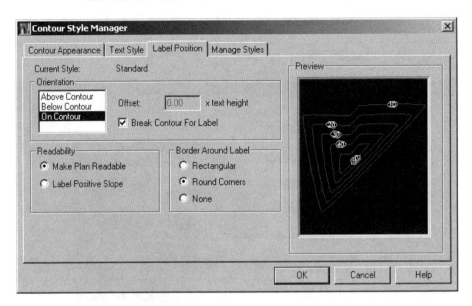

13. Put the label **"On the Contour"**.

14. **Break** the contour for the Label.

15. Select **"Make Plan Readable"**.

16. Choose a border with **"Rounded Corners"** for the label.

17. Select the **Manage Styles** tab.

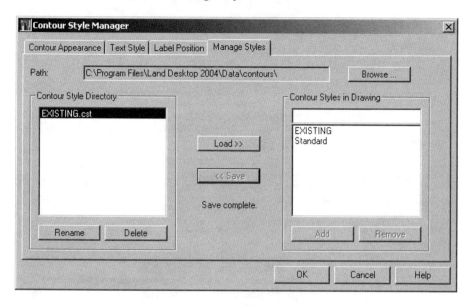

18. With the style named **"Existing"** highlighted click **<<Save>>**.

This saves the new style out to the contour style directory so that it can be loaded and used in other drawings. This gives you a consistent standard that you don't have to recreate with every new drawing.

19. Click **<<OK>>** to see your changes to the selected contour.

Since we changed the name of the contour style the changes will only affect the contour that we selected at the beginning of the exercise. All of the other contours are still using the original contour style named Standard. We need to change their style to Existing so that they will all react consistently to the Existing contour style.

20. Select all of the contours and **Right-click**.

21. From the Right-click menu select ⇒ **Contour Properties**.

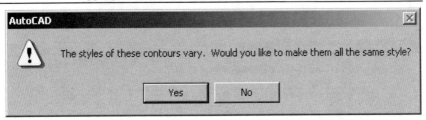

22. You will then see a message that says the contour styles vary and asks if you would like to use the same style for all contours, click **<<YES>>**.

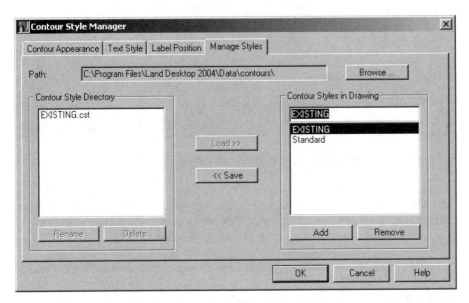

23. In the Contour Style manager select the style **"Existing"** and then click **<<OK>>**.

This will set all of the selected contours to the Existing style. Now that the style has been saved when you create contours in the future you can just select the Existing style in the Create Contours dialog box and the contours will all be created using this style.

5.5   Chapter Summary

In this chapter we created a surface from our survey data. We preformed a slope analysis and created contours. This surface is going to be a critical part of our project because the profiles and cross sections that we create during the design process will all be based on this surface. An accurate surface will allow us to create accurate profiles and cross sections.

Chapter 6

Drafting And Defining Of Proposed Horizontal Alignments

In this chapter we will focus on alignments. We will draft, define, and edit the alignments. Then we will look at what we can do once the alignments are defined into the project database such as stationing and offsets. While some of the drafting utilities that we use in this chapter are nice the real reason that we need the alignments defined into the project is so that we can create profiles and cross sections. We are still in the phase of creating our base design data so that Land Desktop can help us out later on.

- **Preparing a New Design Drawing**

- **Laying Out the Horizontal Alignment**

- **Working With the Alignment Database**

- **Alignment Settings**

- **Stationing and Offsets**

Dataset:

To start this chapter we will be creating a new drawing. You can continue with the project data that you currently have from the end of the previous chapter or you can install the dataset named **"Chapter 6 Cadapult Level 1 Training.zip"** from the CD that came with the book. Extract the dataset to the folder **C:\Land Projects 2004** or whatever folder you have been using as your project path. Extracting the project from the dataset provided will ensure that you have the project and drawings set up correctly for the exercises in the following chapter and overwrite any mistakes that you may have made in previous exercises.

6.1   Preparing A New Drawing For The Design

For the design portion of the project we will use a different drawing. The following exercises will create and setup the design drawing.

6.1.1   Create A New Design Drawing

1.   Select **File** ⇒ **New**.

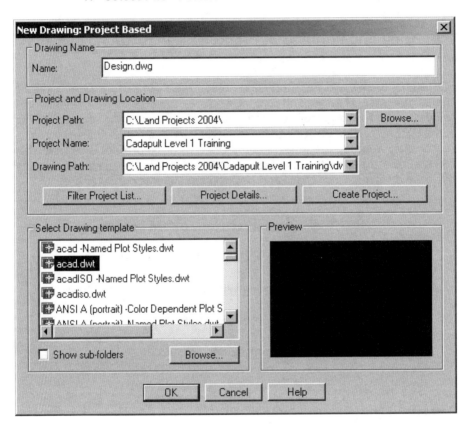

2.   Name the new drawing **"Design"** using the same project **"Cadapult Level 1 Training"** and the **acad.dwt** template.

3.   Click **<<OK>>**.

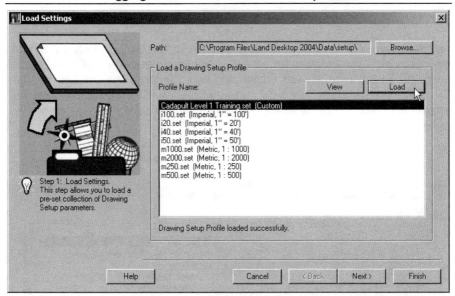

4. Load the **"Cadapult Level 1 Training"** Setup Profile when the first panel of the Drawing Setup wizard appears.

5. Then Click **<<Finish>>** to skip the rest of the wizard.

6.1.2 Setup The Drawing Settings

Since we have not saved out settings to a project prototype in the previous exercises each time we create a new drawing we need to reset the same settings. In this exercise we will set the Point Settings and the Label Settings. We will not go into the detail with these that we have in previous chapters so refer to those exercises for a more detailed explanation of the settings.

1. Select **Projects ⇒ Edit Drawing Settings.**

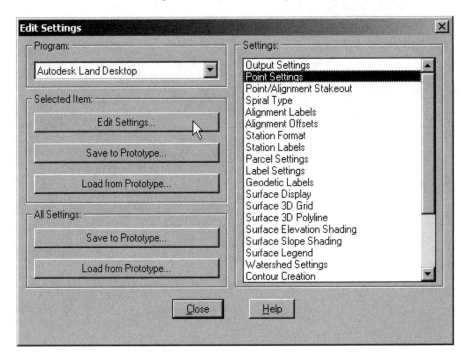

2. Pick **Point Settings** and then select the **<<Edit Setting>>** button. You can also double click on point settings for the same result.

3. Select the **Update** tab of the **Point Settings** dialog box.

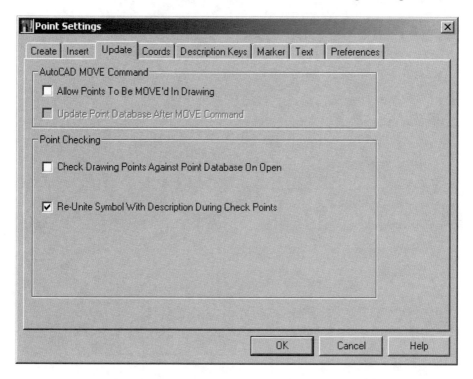

4. **Disable** both options for the **AutoCAD MOVE Command**.

5. Select the **Marker** tab of the **Point Settings** dialog box.

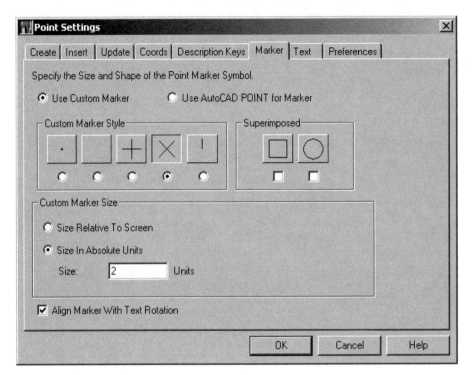

6. Set the **Marker Size** to **2** using **Absolute Units**.

7. Select the **Text** tab of the **Point Settings** dialog box.

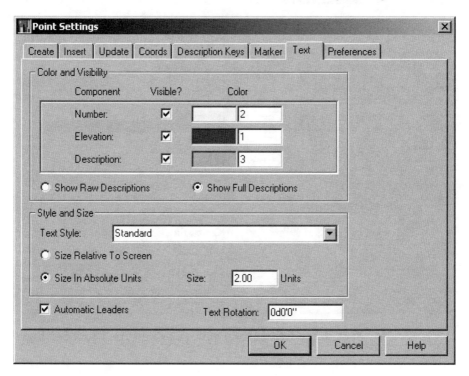

8. Set the **Text Size** to **2** using **Absolute Units**.

9. **Enable** the option to use **Automatic Leaders**.

10. Click **<<OK>>** to save the changes.

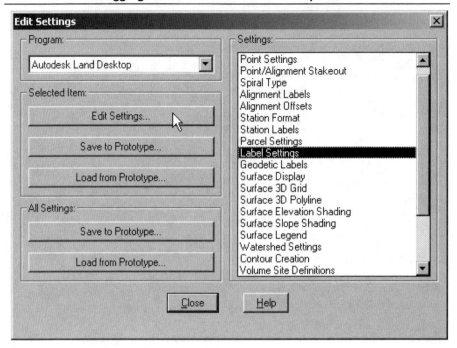

11. Back in the **Edit Setting** dialog box edit the **Label Settings**.

12. Select the **Point Labels** tab of the **Label Settings** dialog box.

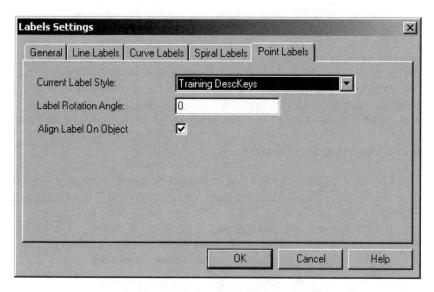

13. Set the current **Point Label Style** to **Training DescKeys**.

14. Click **<<OK>>** to save the changes.

15. Click **<<Close>>** to close the **Edit Settings** dialog box.

6.1.3 Insert The Center Line Points

1. Select **Points ⇒ Insert Points to Drawing**.

2. At the command line type **G** for group.

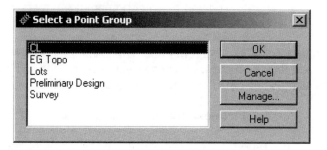

3. Pick the group named **CL** and click **<<OK>>**.

4. **Zoom Extents** to view the points.

6.2   Laying Out The Horizontal Alignment

The horizontal alignment can be laid out by either using AutoCAD commands like Line, Pline, and Fillet or you can use commands from the Land Desktop Lines/Curves pull-down. It doesn't matter which commands you use because they all create standard AutoCAD objects. Later we will define these objects into the Land Desktop Alignment database. That is the point that you must be careful of which commands you are using so that you keep the drawing and the project in sync. For our exercise we will utilize the Land Desktop commands in the Lines/Curves pull-down.

6.2.1   Creating The Tangents

1.   Create a new layer called **T Street CL,** make it **Current** and color **Red.** This layer will contain the T Street centerline.

2.   Select **Lines/Curves** ⇒ **By Point # Range.**

3.   Enter the Point Number Range: **554-558**

4.   **ENTER** to end the command.

5.   Create a new layer called **C Street CL,** make it **Current** and color **Red.** This layer will contain the C Street centerline.

6.   Select **Lines/Curves** ⇒ **Line.**

7.   At the **<<Select point>>** prompt at the command line type **".G"** This will allow you to select the points graphically from the screen to draw your line.

8.   Pick the points for the C Street alignment from left to right. You will be selecting point numbers **564, 563, 562, 561.**

Do not select point 558 and remember this command does not show a rubber band line from the last point as you are drawing so be careful not to pick the same point twice and create a zero length line.

9.   **ENTER** to end the line.

10.   Type **".G"** to turn off the Graphical point selection mode.

11.   **ENTER** to end the command.

6.2.2 Creating The Horizontal Curves

1. Select **Lines/Curves** ⇒ **Curve Between Two Lines**.

You may also use the AutoCAD fillet command.

2. Use the following table of information to place curves between the tangents for the two alignments. The curves in the following table are numbered from left to right.

T Street

Curve #	Type	Value
1	Length	150
2	Radius	200
3	Tangent	50

C Street

Curve #	Type	Value
1	Length	250
2	Tangent	100

6.3   Working With The Alignment Database

Once you are happy with the layout of the alignment geometry you are ready to define the alignment into the Land Desktop project database. Once it is defined as an alignment you need to be careful if you edit the alignment in any way to ensure that the project database and the drawing stay synchronized.

6.3.1   Define The Horizontal Alignments

To define the Horizontal Alignments we will use the same process that we used in Chapter 3 when we defined their preliminary versions.

1. Select **Alignments ⇒ Define From Objects.**

2. **Select a starting point** at the left of the T Street alignment so that it will be defined from left to right.

Be careful to pick the west end of the first tangent in the T Street alignment. The alignment begins with a relatively short tangent and if you select the wrong end it will define the alignment as only that one piece and in the wrong direction. If you do select the starting point incorrectly just Escape to cancel the command and start again.

3. After selecting the starting point of the alignment you can **window the remaining segments.**

4. When asked for a reference point **ENTER** to use the alignment start.

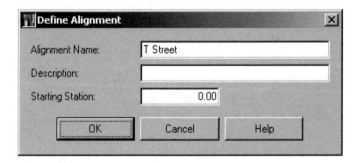

5. Enter an Alignment Name of **"T Street"**.

6. Pick **<<OK>>** to define the alignment.

7. Repeat the process for the **C STREET** alignment.

6.3.2 Editing Alignments

Once the alignments are defined into the Land Desktop project database you have two choices if you need to make changes to them. If you want to make your changes graphically in AutoCAD you must redefine the alignment and overwrite it in the project database. You can also edit the alignment database directly. When using this method the revised alignment will automatically be inserted into your current drawing on the current layer.

1. Set the layer **T Street CL current.**

2. Select **Alignments** ⇒ **Set Current Alignment.**

3. **Enter** to open the alignment librarian.

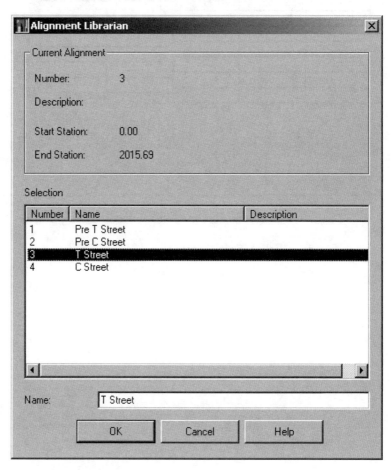

4. Select **T Street.**

5. Select **<<OK>>** to set T Street as the current alignment.

6. Select **Alignments** ⇒ **Edit**.

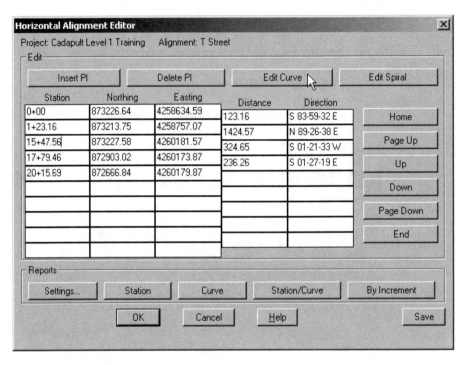

7. Place your cursor in the row containing **Station 15+47.56**.

8. Click **<<Edit Curve>>**.

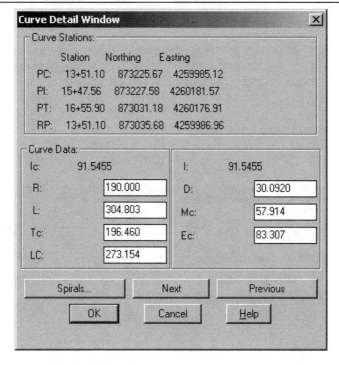

9. Enter a new **radius** of **190**.

10. **ENTER** to apply the new value.

After you have entered the new value you will notice all of the curve information at the top of the dialog box is updated.

11. Click **<<OK>>** to close the Curve Details dialog box.

12. Click **<<OK>>** to close the Alignment Editor.

13. **Save** your changes when prompted.

You will see the alignment flash on the screen as the revised alignment is inserted into the drawing.

6.4   Alignment Settings

Land Desktop has several settings that control the display and layering of alignments and their stationing in your drawing. All of these settings can be saved to a prototype so that they are set automatically when you create a new project.

6.4.1   Station Display Format

1.   Select **Alignments** ⇒ **Station Display Format**

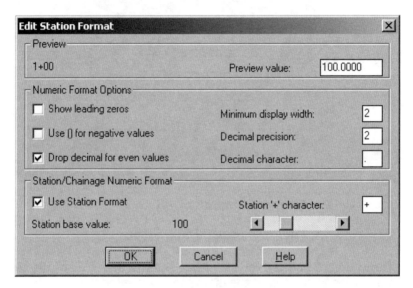

Here you control the numeric display of your Stationing. These settings can be saved to your project prototype.

2.   Make any desired changes. You will see the effects of those changes in the preview at the top of the dialog box.

3.   Click <<**OK**>> to save your changes.

6.4.2   Alignment Labels

1.   Select **Alignments** ⇒ **Alignment Labels**.

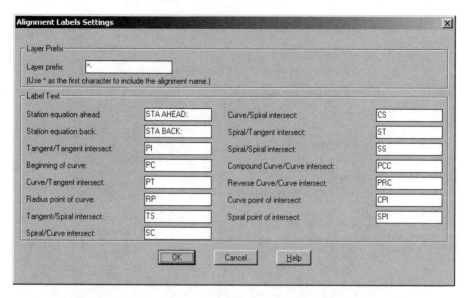

This dialog box gives you control over how your alignment is labeled. It also allows you to set up a Layer Prefix for all layers associated with alignments. An asterisk can be used to add the alignment name as a layer prefix. This layer prefix is very important when you have more than one alignment displayed in your drawing. These settings can also be saved to a project prototype.

2.   Enter **"*-"** as the layer prefix to use the alignment name as a prefix to all of the alignment layers.

3.   Click **<<OK>>** to save your changes.

6.4.3 Station Label Settings

1. Select **Alignments** ⇒ **Station Label Settings**.

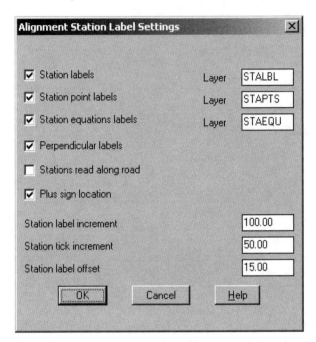

This dialog box allows you to control the actual positions of the station labels relative to the alignment. If you decide to use "Perpendicular Labels" you should also select the "Plus Sign Location" for best results.

2. **Enable** the option for **Perpendicular labels**.

3. **Enable** the option for **Plus sign location**.

4. Click **<<OK>>** to save your changes.

6.5   Stationing And Offsets

Once an alignment is defined into the Land Desktop project database you can station the alignment and create offsets based on the data stored in the project.

6.5.1   Create Stationing In The Drawing

1. Select **Alignments ⇒ Set Current Alignment.**

2. **Enter** to open the alignment librarian.

3. Select **T Street.**

4. Select **Alignments ⇒ Create Station Labels.**

5. Accept the defaults for the beginning and ending station.

6. Repeat the process for C Street.

Land Desktop uses the current text style when creating stations. If you want to change the display of your stationing you must first change the appropriate settings as we did in the previous exercises, then recreate the station labels. When you create station labels you will be given the option at the command line to delete the existing stationing layers.

6.5.2 Creating Alignment Offsets

Once an Alignment as been defined into the project and any necessary editing has been completed you can create offsets from that alignment. You can create up to four offsets on each side of the alignment at a time and you have the option to define them all as alignments.

1. Select **Alignments ⇒ Set Current Alignment.**

2. **Enter** to open the alignment librarian.

3. Select **T Street.**

4. Select **Alignments ⇒ Create Offsets.**

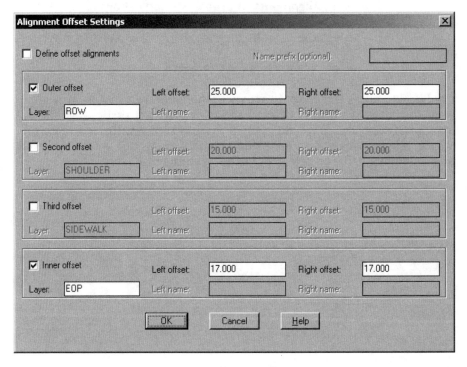

5. Enter a **25'** offset for the right of way

6. Enter a **17'** offset for the edge of pavement.

7. Click **<<OK>>** to create the offsets.

8. Repeat the process for C Street.

9. When the offsets were created they also created two new layers for each alignment. The layers are created with a color of White. Change the layer colors as shown below.

C STREET-EOP	Yellow
C STREET-ROW	Green
T STREET-EOP	Yellow
T STREET-ROW	Green

6.6 Chapter Summary

In this chapter we created our alignments and defined them into the project database. We also looked at the importance of keeping the objects that represent the alignments in the drawing, and definition of the alignments in the project synchronized. The main key to this synchronization is remembering to redefine the alignment if you make any changes in the drawing. If you make changes in the database through the Edit Alignment command it will automatically insert the revised alignment into the drawing. This has prepared us for the next two chapters where we will combine the alignments with the existing ground surface to create profiles and cross sections.

Chapter 7

Working With Profiles

Now that we have a horizontal alignment and a surface both defined into the Land Desktop project we are ready to create a profile of the existing ground along T Street. Creating a profile of an existing surface is a step-by-step process that takes the horizontal information from the alignment and the vertical information from the surface and combines them to generate the profile.

Once a profile of the existing ground has been created we will layout and define a Finished Ground Profile. This process will be very similar to the way that we laid out and defined the horizontal alignment.

- **Profile Settings**

- **Creating The Existing Ground Profile**

- **Profiles of Finished Ground**

Dataset:

To start this chapter we will continue working in the drawing named **Design.dwg.** You can continue with the drawing and project data that you currently have from the end of the previous chapter or you can install the dataset named **"Chapter 7 Cadapult Level 1 Training.zip"** from the CD that came with the book. Extract the dataset to the folder **C:\Land Projects 2004** or whatever folder you have been using as your project path. Extracting the project from the dataset provided will ensure that you have the project and drawings set up correctly for the exercises in the following chapter and overwrite any mistakes that you may have made in previous exercises.

7.1   Profile Settings

As with almost everything in Land Desktop, Profiles have settings that control everything from layers, to labels, and accuracy. Changes to these settings that you would like to be permanent should be saved to a prototype so that you do not have to reset your standard with every new drawing.

In the following exercises we will access the Profile Settings from the Profiles pull-down. However, these same settings can be changed with the Edit Drawing Settings command found in the Projects pull-down menu that we have used in previous exercises.

7.1.1   Profile Label Settings

1. Open the drawing **Design.dwg** from the project **Cadapult Level 1 Training** if it is not already open.

2. Switch to the Civil Design Menu Palette.   Select **Projects** ⇒ **Menu Palettes**.

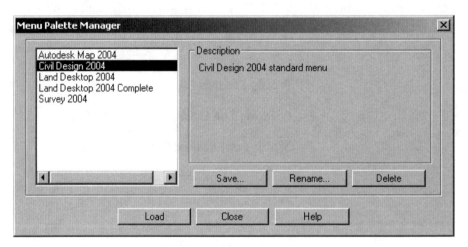

This will load the Civil Design pull-down menus including the Profiles pull-down.

3. Select **Profile** ⇒ **Profile Settings** ⇒ **Labels and Prefix**.

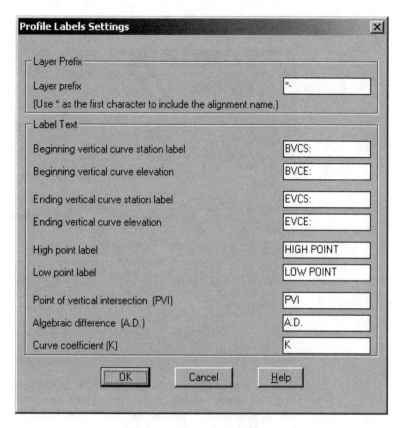

4. For the Layer Prefix enter **"*-"** this will use the alignment name as a prefix for all of the profile layers.

5. Click **<<OK>>** to save your changes.

7.1.2 Existing Ground Profile Layer Settings

1. Select **Profile** ⇒ **Profile Settings** ⇒ **EG Layers**.

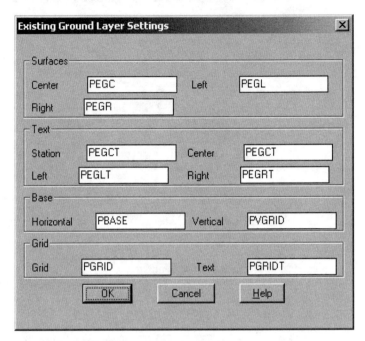

2. Review the Layer Names for all of the Existing Ground Profile Layers. Remember they will all have the Alignment Name as a Prefix because of the "*" we entered in the Profile Label Settings dialog box.

3. Click **<<OK>>** to save any changes.

7.1.3   Finished Ground Profile Layer Settings

1. Select **Profile** ⇒ **Profile Settings** ⇒ **FG Layers.**

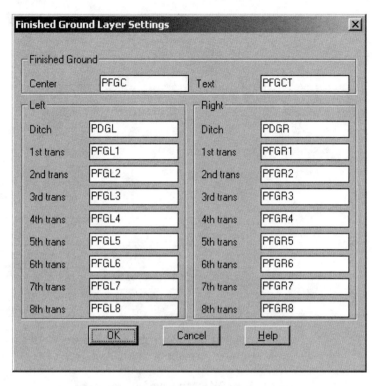

2. Review the Layer Names for all of the Finished Ground Profile Layers. Remember they will all have the Alignment Name as a Prefix because of the "*" we entered in the Profile Label Settings dialog box.

3. Click **<<OK>>** to save any changes.

7.1.4 Profile Value Settings

1. Select **Profile** ⇒ **Profile Settings** ⇒ **Values**.

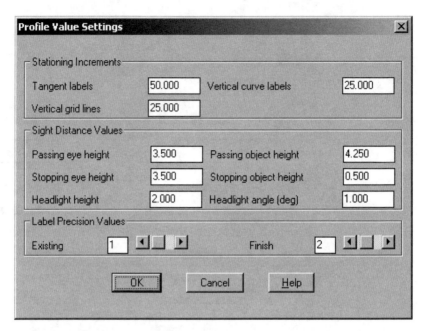

2. This dialog box controls your Profile Values including Stationing increment and Label Precision of the Profile.

3. Click **<<OK>>** to save any changes.

7.1.5   Profile Sampling Settings

1. Select **Profile** ⇒ **Profile Settings** ⇒ **Sampling.**

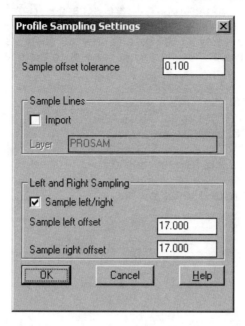

2. **Enable** the option to **Sample Left and Right** profiles.

3. Set the left and right sample offset to **17.**

4. Click **<<OK>>** to save any changes.

7.2   Creating The Existing Ground Profile

Creating an Existing Ground Profile from a surface is one of the few step-by-step processes in Land Desktop that rarely changes from project to project. Here you really get a chance to take advantage of all the work that you have done creating all the data in previous chapters. Below are the steps that we will use. You may want to refer to them later for reference.

1.   Set the current alignment.

2.   Set the current surface.

3.   Sample from the surface.

4.   Create the full profile.

7.2.1   Sampling The Profile

1. Set the current alignment. **Alignments** ⇒ **Set Current Alignment.**

2. Pick the alignment on the screen or Return to bring up the Alignment Librarian and pick **"T Street"**.

3. Set the current surface. **Profiles** ⇒ **Surfaces** ⇒ **Set Current Surface.**

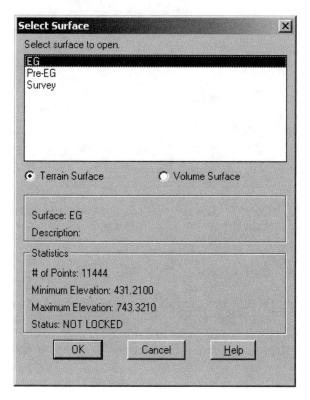

4. Select **"EG"**.

5. Click **<<OK>>**.

6.   Sample from the surface. **Profiles** ⇒ **Existing Ground** ⇒
     Sample from Surface.

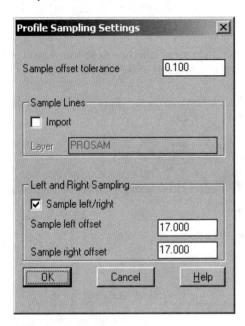

7.   Confirm that the settings are enabled to sample left and
     right offsets at 17'. This was previously setup in the Profile
     Settings exercise.

8.   Click **<<OK>>**.

9.   Accept the **default** for the **beginning and ending stations**
     when prompted at the command line.

Now the existing ground profile information has been sampled and saved in
your project.

7.2.2   Setting The Profile Scale

The drawing scale, as set in the Drawing Setup command, controls the vertical exaggeration of your profile.

1.   Select **Projects** ⇒ **Drawing Setup**.

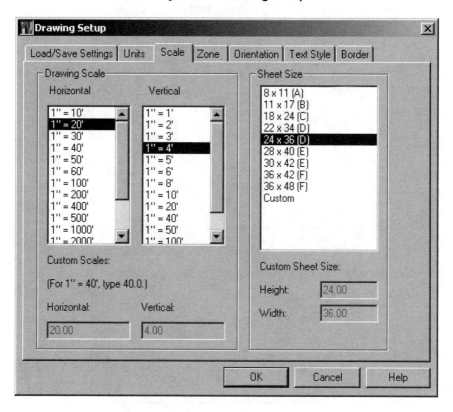

2.   Select the **Scale** tab.

3.   Adjust the Horizontal or Vertical scale as desired to change the vertical exaggeration of your profile. For our exercise set the horizontal scale to **1"=20'** and the vertical scale to **1"=4'**. This will give our profile a 5:1 exaggeration.

4.   Click **<<OK>>** to save your changes.

7.2.3 Drawing The Profile

Since the profile data has been sampled you can draft the profile in any drawing that is associated with this project at any time. You do not need to resample unless there is a change to the existing ground surface.

1. Select **Profiles** ⇒ **Create Profile** ⇒ **Full Profile.**

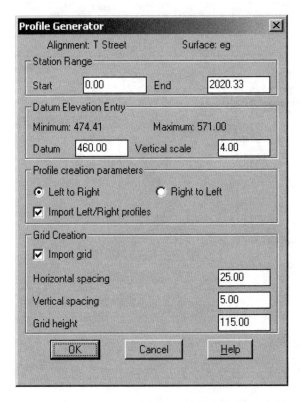

2. **Enable** the option to **Import Left/ Right profiles.**

3. **Enable** the option to **Import the Grid.**

4. Use the default values for the Grid Spacing and Height.

5. Click **<<OK>>.**

6. At the command line you will be asked for a Starting Point for the profile. You can either pick a point on the screen or type in an X,Y coordinate. For our exercise we will type in a coordinate. The starting point will define the lower left corner of the profile. At the command line enter **4261100,872800.**

7. Answer **"Yes"** to **Delete existing profile layers.**

8. **Zoom Extents** to view the profile.

Note: Once you have placed the profile it is important not to move it. LDT references this profile and its finished ground centerline from the point that you pick as the starting point for the profile. If you must move the profile you will need to register its new position with Land Desktop or recreate the profile in the desired position.

9. Change the following layer colors:

T STREET-PEGC	- Red
T STREET-PEGCT	- Red
T STREET-PEGL	- Yellow
T STREET-PEGLT	- Yellow
T STREET-PEGR	- Green
T STREET-PEGRT	- Green
T STREET-PGRID	- 8

7.3   Profiles of Finished Ground

You will find that laying out the Finished Ground or Proposed Profile is very similar to the process of working with Horizontal Alignments. You start by laying it out graphically in the drawing. Then you define it into the project database. Once it is defined into the project you need to be careful to keep the data synchronized between the drawing and the project if you make any changes to the profile.

In the following exercises we will take some extra steps to illustrate the process of keeping the drawing and the project in sync. We will define and view the profile in the Vertical Alignment Editor several times so that you can see where the changes are being saved during the editing process. For these reasons of illustration there may be extra steps in the exercises that you may not need in a typical project.

7.3.1   Construct Finished Ground Centerline

The existing ground profile is now completed. We are ready to construct the Finished Ground Centerline. This can be drawn in our profile by using the LDT create tangents command or by just sketching line segment with the AutoCAD Line command. If you want to use the Line command you must set the finished ground centerline (PFGC) layer current. You can do this with the command **Profiles ⇒ FG Centerline Tangents ⇒ Set Current Layer**. In our exercise we will use the Land Desktop "Create Tangents" command. It automatically puts our tangents on the correct layer as well as giving us options for Station, Elevation, and Grade that take into account the vertical exaggeration of the profile.

1.   Zoom Into the profile if you have not already done so.

2.   Select **Profiles ⇒ FG Centerline Tangents ⇒ Create Tangents.**

3.   Type **S** for Station. Accept the default beginning station of **0**. Accept the **default** beginning elevation of **474.92**.

4.   Type **L** for Length. Enter a length of **800**. Type **G** for grade and use a grade of **5.25**.

5.   For the next point use a Station of **1040** and a Grade of **-7**.

6.   Next use a Station of **1585** and an Elevation of **561**.

7.   Finally use an **Endpoint Osnap** and snap to the end of the existing ground centerline.

8. **Enter** to end the command.

At this point the vertical alignment has only been sketched into the drawing and has not been defined in the project database. Once it has been defined into the project database you can extract information from it and create reports. You can also edit the vertical alignment in a spreadsheet style dialog box similar to the way that we edited the horizontal alignment. If you make any changes to the vertical alignment graphically then you will need to redefine it into the project. If you make any changes in the vertical alignment editor you will be asked if you want to import them into the drawing or you can import them at any time by using the command **Profiles ⇒ FG Vertical Alignments ⇒ Import**.

7.3.2 Defining And Editing The Vertical Alignment

1. Select **Profiles ⇒ FG Vertical Alignments ⇒ Define FG Centerline**.

The finished ground centerline layer will temporarily be isolated for you.

2. Pick the beginning point of the proposed profile then select the entire profile with a crossing window. **ENTER** to finish the selection.

The Finished Ground Centerline profile is now defined in to the project.

3. Select **Profiles ⇒ Edit Vertical Alignments**.

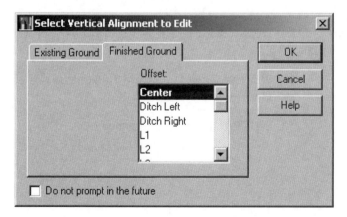

4. Select the **"Center"** offset on the **Finished Ground** Tab.

5. Click **<<OK>>** to open the Vertical Alignment Editor.

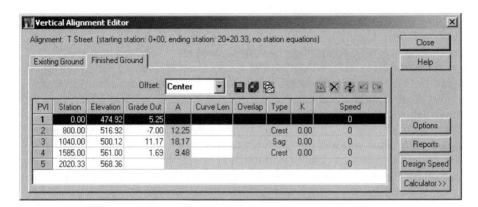

In the Vertical Alignment Editor you can see the PVI (Point of Vertical Intersection) Stations, Elevations, and the grades between them. You can also make changes to any of this information or add vertical curves in this dialog box. However, for our exercise we will make some of these changes graphically first.

6. Click **<<Close>>** to leave the Vertical Alignment Editor without making any changes.

7. Isolate the vertical centerline layer:
   T STREET-PFGC

8. Select **Profiles** ⇒ **FG Centerline Tangents** ⇒ **Change Grade 2.**

The Change Grade 1 and Change Grade 2 commands move a PVI in between 2 selected tangents by changing the grade of the first or second tangent respectively.

9. Pick the second and third Tangents in the alignment and enter a **New Grade Out** of **10.**

10. **ENTER** to end the command.

Now we will add vertical curves to the alignment.

11. Select **Profiles** ⇒ **FG Vertical Curves.**

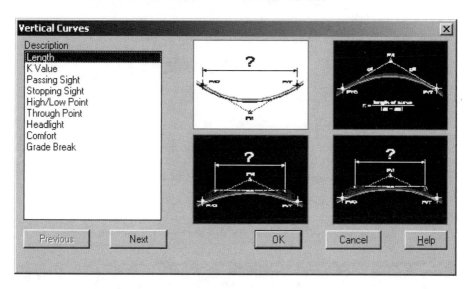

12. Define the curves by **length**.

13. Use the following curve lengths:

Curve 1	100
Curve 2	300
Curve 3	250

14. **ENTER** to end the command.

Now that we have updated the profile graphically we nee to redefine the Vertical Alignment.

15. Select **Profiles** ⇒ **FG Vertical Alignments** ⇒ **Define FG Centerline.**

16. Pick the beginning point of the proposed profile then select the entire profile with a crossing window. **ENTER** to finish the selection.

17. Select **Profiles** ⟹ **Edit Vertical Alignments**.

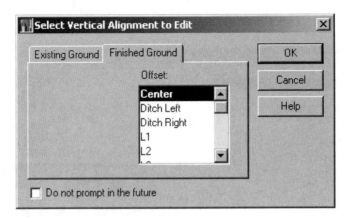

18. Select the **"Center"** offset on the **Finished Ground** Tab.

19. Click **<<OK>>** to open the Vertical Alignment Editor.

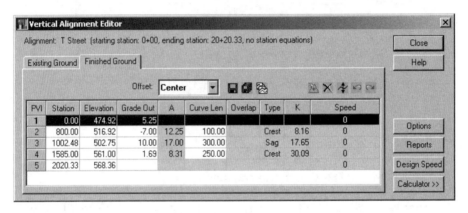

You should see that our profile now has data for the vertical curves and the Grade Out of PVI #3 has been changed to 10.

20. Click the **<<Calculator>>** button.

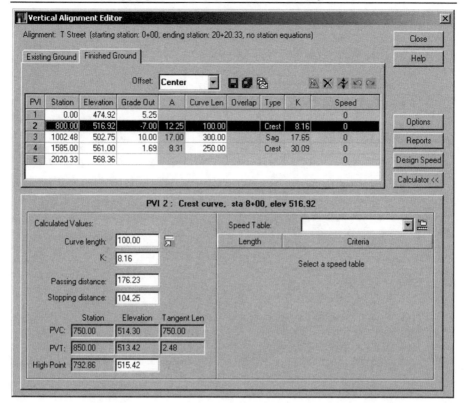

21. Now select the row containing PVI number 2 and the first vertical curve in the alignment.

The Curve Calculator at the bottom of the dialog box will display all of the curve geometry and allow you to check multiple scenarios before applying the changes to the profile.

22. Change the **Curve Length** to **150**.

After changing the curve length to 150 you will notice that there are red check marks in the Overlap column of the table. This is showing that the vertical curves at PVI 2 and PVI 3 are too big and overlap one another.

23. Change the **Length of Curve #2** to **250**.

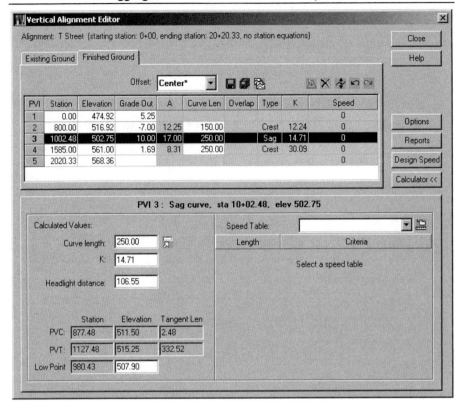

This change should resolve the problem of the overlap.

>   24. Click **<<Close>>** to leave the Vertical Alignment Editor.

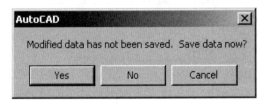

>   25. **Save** your changes.

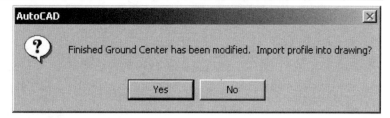

26. **Import the new Profile.**

When prompted at the command line answer **Yes** to:

27. **"Label tangents and vertical curves"**

28. **"Delete finished ground profile layer"**

29. **"Turn ON** all of the **T Street** layers".

After reviewing the design of the profile you may notice that there is a deep fill around station 12+00. To improve this we can use the Vertical Alignment Editor to move PVI 3 and 4. This will hold the grade between these PVI's that is already at the maximum allowed for our project.

30. Select **Profiles** ⇒ **Edit Vertical Alignments**

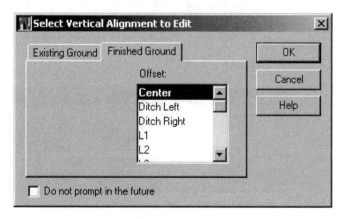

31. Select the **"Center"** offset on the **Finished Ground** Tab.

32. Click **<<OK>>** to open the Vertical Alignment Editor.

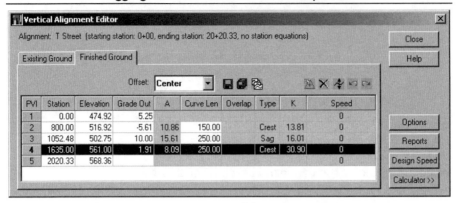

33. Change the **Station** of **PVI 3** to **1052.48**

34. Change the **Station** of **PVI 4** to **1635.00**

This shifts the two PVI's 50' and holds the 10% grade between them.

35. Click **<<Close>>** to leave the Vertical Alignment Editor.

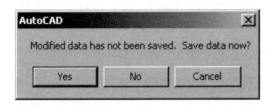

36. **Save** your changes.

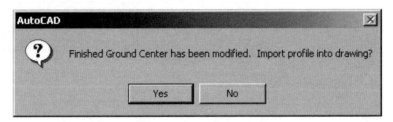

37. **Import the new Profile.**

When prompted at the command line answer **Yes** to:

38. "**Label tangents and vertical curves**".

39. "**Delete finished ground profile layer**".

The vertical alignment has now been designed and defined into our project. It will control our template when we sample cross sections and setup the Design Control.

7.4   Chapter Summary

In this chapter we have created both an existing and finished ground profile. They are based on the horizontal alignment and the existing ground surface that we defined in the previous chapters. So you are finally getting to see some of the payoff of all the work that was done to carefully set up the project and compile all of the base data. This theme will continue because in the next chapter the finished ground profile that we just created will become the base data that controls the position of the road template and the design of our cross sections.

Chapter **8**

Templates And Cross Sections

In this chapter you will define and edit a road template that will be applied to Cross Sections that will be sampled. The template will then be applied to the cross sections through the Design Control commands. Finally, you can extract data from your design to create points, surfaces, and volume calculations.

It is also important to think about Templates and Cross Sections as just another set of tools in Land Desktop. Don't limit yourself to thinking of them as only useful for a road design. They can be a grading tool for projects such as ditches, berms, or bio-swales. Once you understand how these tools work it will open up many different areas that they can make you more productive and efficient.

- **Drawing the Template**
- **Defining And Editing Templates**
- **Template Volumes**
- **A Grid Of 3D Faces**
- **Material And Point Code Tables**
- **Sections And Design Control**
- **Setting Points From A Template**
- **Road Output**

Dataset:

To start this chapter we will continue working in the drawing named **Design.dwg**. You can continue with the drawing and project data that you currently have from the end of the previous chapter or you can install the dataset named **"Chapter 8 Cadapult Level 1 Training.zip"** from the CD that came with the book. Extract the dataset to the folder **C:\Land Projects 2004** or whatever folder you have been using as your project path. Extracting the project from the dataset provided will ensure that you have the project and drawings set up correctly for the exercises in the following chapter and overwrite any mistakes that you may have made in previous exercises.

8.1   Drawing the Template

When drawing a template you have the choice of using the AutoCAD **Polyline** command or the LDT command **Cross Sections** ⇒ **Draw Template.** In each command you must draw closed polylines. The difference is that the Draw Template command compensates for your vertical drawing scale that is set in your drawing setup. If you use the Polyline command must manually compensate for the vertical scale.
The template that we are using is symmetrical so we only need to draw the left side. LDT will mirror it to create the right side when you define the template.

For our exercise we will use the Land Desktop Draw Template command. Before we start make sure that you have the Civil Design menu palette loaded so that you have the Cross Sections pull-down available. You will also want to zoom in to a small blank area in the drawing. You will be able to use the transparent pan and zoom commands while you are drawing the template, so don't worry too much about the exact size and area that you have displayed at this point.

Finally, turn off any running Object Snaps; they can give you some unexpected results during the Draw Template command.

Below is a graphic that shows the template that we will be drawing for our project.

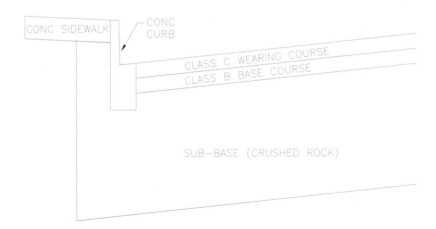

8.1.1 Drawing The Wearing Course

1. Open the drawing **Design.dwg** from the project **Cadapult Level 1 Training** if it is not already open.

2. Confirm that **Object Snaps** are turned **Off**.

3. Select **Cross Sections** ⇒ **Draw Template**.

4. Pick a point in the upper middle of your screen.

5. Type **G** for Grade and -2 for 2%. Use an offset of -**16 feet**.

6. Next use **Relative** with an **Offset** of **0** and a Change in Elevation of **-0.167**.

7. Now a Grade of **2** and a change in offset of **16**.

8. **ENTER** to end the line. Since this is a symmetrical template when the left side is mirrored it will close the polygon.

8.1.2 Drawing The Base Course

1. The Draw Template Command should still be active. If it is not, select **Cross Sections** ⇒ **Draw Template**. For the starting point of the next surface in the template use the **End Object Snap** to select the last point on the asphalt surface.

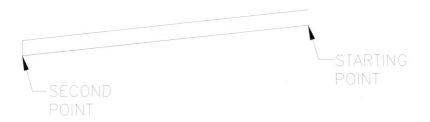

2. Starting point: **end**

3. Grade (%) [Relative/Slope/Points/Close/Undo/eXit]: **P**

4. Select point (Relative/Grade/Slope/Close/Undo/eXit): **end**

5. Select point (Relative/Grade/Slope/Close/Undo/eXit): **R**

6. Change in offset (Grade/Slope/Close/Points/Undo/eXit): **0**

7. Change in elev: **-.167**

8. Change in offset (Grade/Slope/Close/Points/Undo/eXit): **G**

9. Grade (%) (Relative/Slope/Points/Close/Undo/eXit): **2**

10. Change in offset: **16**

11. **Enter** to end the line.

8.1.3   Drawing The Curb

1.   Begin the curb at the top left edge of pavement.

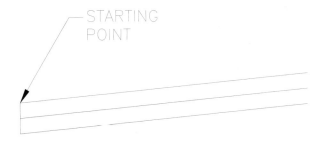

2.   Starting point: **end**

3.   Grade (%) (Relative/Slope/Points/Close/Undo/eXit): **-2**

4.   Change in offset: **-.95**

5.   Grade (%) (Relative/Slope/Points/Close/Undo/eXit): **R**

6.   Change in offset (Grade/Slope/Close/Points/Undo/eXit): **-.05**

7.   Change in elev: **.5**

8.   Change in offset (Grade/Slope/Close/Points/Undo/eXit): **-.5**

9.   Change in elev: **0**

10. Change in offset (Grade/Slope/Close/Points/Undo/eXit): **0**

11. Change in elev: **-1**

12. Change in offset (Grade/Slope/Close/Points/Undo/eXit): **1.5**

13. Change in elev: **0**

14. Change in offset [Grade/Slope/Close/Points/Undo/eXit]: **P**

15. Select point [Relative/Grade/Slope/Close/Undo/eXit]: **end**

Pick the endpoint at the bottom left corner of the base course.

16. Select point [Relative/Grade/Slope/Close/Undo/eXit]: **C**

This closes and ends the Curb surface.

8.1.4   Drawing The Sidewalk

1. Begin the Sidewalk at the back of curb using an endpoint snap.

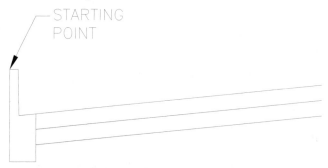

2. Starting point: **end**

3. Select point [Relative/Grade/Slope/Close/Undo/eXit]: **G**

4. Grade (%) [Relative/Slope/Points/Close/Undo/eXit]: **1**

5. Change in offset: **-5**

6. Grade (%) [Relative/Slope/Points/Close/Undo/eXit]: **R**

7. Change in offset [Grade/Slope/Close/Points/Undo/eXit]: **0**

8. Change in elev: **-.25**

9. Change in offset [Grade/Slope/Close/Points/Undo/eXit]: **G**

10. Grade (%) [Relative/Slope/Points/Close/Undo/eXit]: **-1**

11. Change in offset: **5**

12. Grade (%) [Relative/Slope/Points/Close/Undo/eXit]: **C**

This closes and ends the Sidewalk surface.

8.1.5   Drawing The Sub-base

1. Finally start the Sub-base at the bottom centerline of the Base

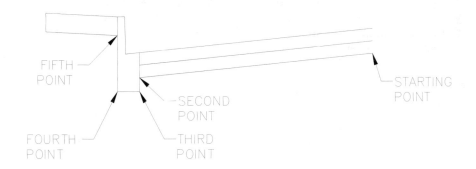

2. Starting point: **end**

3. Grade (%) [Relative/Slope/Points/Close/Undo/eXit]: **P**

4. Select point (Relative/Grade/Slope/Close/Undo/eXit): **end**

5. Select point (Relative/Grade/Slope/Close/Undo/eXit): **end**

6. Select point (Relative/Grade/Slope/Close/Undo/eXit): **end**

7. Select point (Relative/Grade/Slope/Close/Undo/eXit): **end**

8. Select point (Relative/Grade/Slope/Close/Undo/eXit): **G**

9. Grade (%) (Relative/Slope/Points/Close/Undo/eXit): **1**

10. Change in offset: **-2**

11. Grade (%) (Relative/Slope/Points/Close/Undo/eXit): **R**

12. Change in offset (Grade/Slope/Close/Points/Undo/eXit): **0**

13. Change in elev: **-2**

14. Change in offset (Grade/Slope/Close/Points/Undo/eXit): **G**

15. Grade (%) (Relative/Slope/Points/Close/Undo/eXit): **2**

16. Change in offset: **19.5**

17. **Enter** twice to end the line and the Draw Template command.

The completed template should look like the figure below. At this point your template is drawn correctly. However, it has not been defined into the Land Desktop template library. Currently it is only a collection of polylines in the drawing.

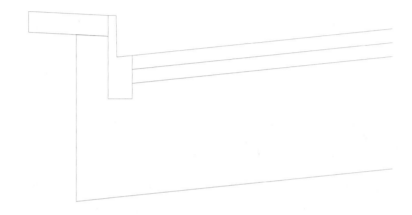

8.2   Material And Point Code Tables

Material and Point code tables are lists of data that can be used to add information to your template. The Material table will be used during the Define Template command and the Point Code table will be used during the Edit Template command. So it is important that they are defined before you attempt to define or edit the template. Both of these tables are stored in the template library so you can use them with any project once you have defined them the first time.

8.2.1   Creating a Material Table

Material Tables are lists of the materials that make up the surfaces in your template. These materials do not have any properties associated with them such as weight or density; they are simply tags for when you calculate template volumes. The material names that you assign to the surfaces in the template will be shown in the summary of the template surface volume calculation.

1.   Select **Cross Sections** ⇒ **Templates** ⇒ **Edit Material Table.**

2.   Click **<<New>>** to create a New Table.

3. Name the table **"Level 1 Training"**

4. You are automatically prompted to enter the first material name. The first name is **"Wearing Course."**

5. Click **<<New>>** in the Materials section and enter the following material names:

    Base Course
    Crushed Rock
    Conc Curb
    Conc Sidewalk

6. Click **<<OK>>** to save the new Material Table and close the dialog box.

8.2.2   Creating a Point Code Table

Point Codes are used to define specific points on a template. These Point Codes when used in conjunction with Description Keys give you an automated way to set points in your drawing after a template has been applied to an alignment. Defining Point Codes and Point Codes Tables is very similar to defining Material Tables.

1. Select **Cross Sections** ⇒ **Templates** ⇒ **Edit Point Code Table**.

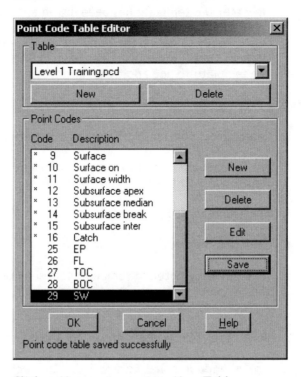

2. Click **<<New>>** to create a New Table.

3. Name the table **"Level 1 Training"**

4. Click **<<New>>** button in the Point Code section and add the following codes:

25	**EP**
26	**FL**
27	**TOC**
28	**BOC**
29	**SW**

    The first 16 point codes are predefined and you should not try to edit them. Codes 17 through 24 are reserved for future use by Autodesk. So the first point code that you can define is number 25.

5. Click **<<OK>>** to save the new Material Table and close the dialog box.

8.3 Defining And Editing Templates

Now that the template has been drawn and the material table and point code tables have been defined, we are finally ready to define the template into the template library. Once the template has been defined we will edit the template to add additional information.

8.3.1 Defining a Template

1. Select **Cross Sections** ⇒ **Templates** ⇒ **Define Template**.

2. Use an endpoint snap to select the **Finished Ground Reference Point** at the Crown of the road.

3. The Template is **Symmetrical**.

4. Select all **5** objects in the template.

5. After pressing **Enter** to complete your selection set one of the template surfaces should be highlighted. All of the surfaces are Normal.

6. Select the appropriate material for the highlighted surface.

7. Use an **endpoint** snap to select the **Connection Point Out** at the top, back of the Sidewalk.

8. Use **Datum** Number **1**.

9. Pick the Datum Points starting at the top, back of sidewalk using your object snaps. Then continue to select each point around the bottom of the template as shown by the thick dashed line below.

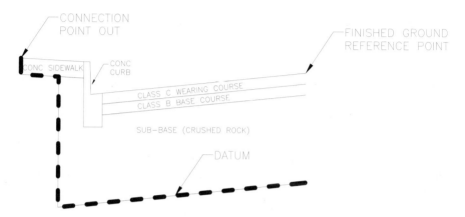

10. **ENTER** when you are finished selecting the datum points.

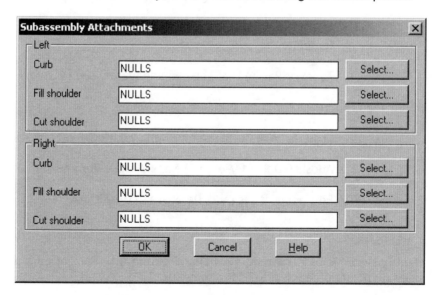

11. We will not be using subassemblies with this template so leave them set to null and click **<<OK>>**.

12. Save the Template as **"Level 1"**

13. Answer **No** when asked if you would like to define another template.

14. You may now erase the template lines on your screen.

The template has been defined to the Land Desktop template library.

8.3.2   Editing The Template To Add Point Codes

In our last exercise we defined the template. However, we did not apply our Point Codes that we set up in the Point Code Table. Point Codes are not a required part of the template so the template must be edited to add them. The Edit Template command also gives you a chance to change any Material Names, edit the Datum, create a Top Surface, add Transition Regions and much more.

For this exercise we will add Point Codes. The Edit Template command is completely driven by the command line. So it is important that you watch the command line and follow the prompts.

1. Select **Cross Sections** ⇒ **Templates** ⇒ **Edit Template**.

2. Select **"Level 1"** and click **<<OK>>**.

3. Pick a point in the middle of the screen to import the template.

4. Zoom or Pan so that you can adequately see the template.

Follow the prompts at the command line as shown below:

5. Edsrf/SAve/eXit/ASsembly/Display/SRfcon/Redraw <eXit>: **E**

6. Addsurf/Delsurf/Modify/MName/Points/Redraw/eXit <eXit>: **P**

7. Edit point codes (Add/Delete/eXit) <eXit>: **A**

8. Select point code location: **end**

Identify the Edge of Pavement, Flow Line, Top of Curb, Back of Curb and Sidewalk points **on each side** of the template.

9. When you are finished selecting locations for Point Codes **"Enter"** to return to the Edit Point Code prompt.

10. Do not exit the Edit Template command.

8.3.3   Editing The Template To Add The Top Surface

The Top Surface of a template is exactly what the name implies; it is a surface defining the top of the template. It can be used later to create points for staking, to create 3D faces for visualization, and for surface creation. You can define up to 8 top surfaces on a single template, which can allow you to use the same template for many different design scenarios. Similar to Point Codes, you are not required to define a Top Surface for every template, so you can determine the level of detail that you need in your template and define it accordingly.

1. Continue to follow the prompts at the command line in the Edit Template command as shown below. (If you are not currently in the Edit Template command start it now by selecting **Cross Sections** ⇒ **Templates** ⇒ **Edit Templates.**)

2. Edit point codes (Add/Delete/eXit) <eXit>: **X**

3. Addsurf/Delsurf/Modify/MName/Points/Redraw/eXit <eXit>: **X**

4. Edsrf/SAve/eXit/ASsembly/Display/SRfcon/Redraw <eXit>: **SR**

5. Connect/Datum/Redraw/Super/Topsurf/TRansition/eXit <eXit>: **T**

6. Top surface number <1>: **1**

7. Pick top surface points (left to right): **end**

Draw the Top surface by snapping to each point across the top of the template from the left back of sidewalk to the right back of sidewalk.

8. **ENTER** when you are finished selecting the top surface points.

9. Type **X** to **exit the Edit Template** command.

10. **Save** and **Overwrite** the **Level 1**template.

11. **Erase** the template geometry from the screen.

8.4   Sections And Design Control

We need to sample Cross Sections along our alignment so that we can use Design Control later to apply our template to these sections. This is similar, in many ways, to doing a Cross Section design by hand. Our template will be applied and will daylight perfectly at each location that we sample a section. So it is important that we sample sections at any critical geometry points along the alignment. In short the more sections you sample the more accurate your design will be because in between each section Land Desktop will just be transitioning from one section to the next.

8.4.1   Sampling Sections

1. Set the current alignment. **Alignments ⇒ Set Current Alignment.** Pick the alignment on the screen or Return to bring up the Alignment Librarian and pick "**T Street.**"

2. Set the current surface. **Cross Sections ⇒ Surfaces ⇒ Set Current Surface.**

3. Set surface "**EG**" current.

4. Sample from the surface. **Cross Sections** ⇒ **Existing Ground** ⇒ **Sample from Surface**. Fill out the dialog box as follows.

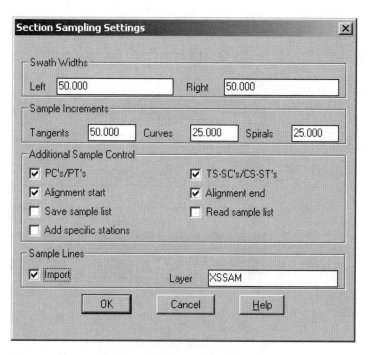

5. Click **<<OK>>** to begin sampling.

6. Accept the default Beginning and Ending stations.

8.4.2 Design Control

Now we are ready to apply our template through Design Control. We will use our **"Level 1"** template and daylighting with a 3:1 slope for the entire alignment.

1. Select **Cross Sections** ⇒ **Design Control** ⇒ **Edit Design Control**.

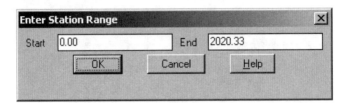

2. Accept the default station range for the entire alignment.

3. Click **<<Template Control>>**.

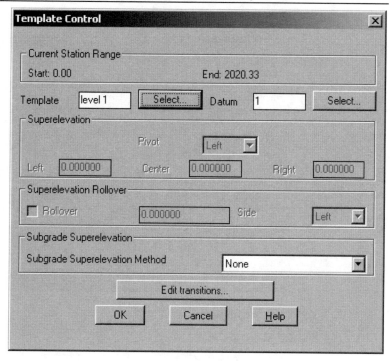

4. Select the **"Level 1"** template.

5. Click **<<OK>>** to save your changes.

6. Click <<**Slopes>>**.

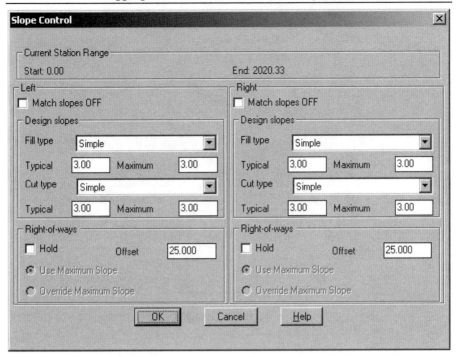

7. Enter **3:1** Slopes for the left and right cut and fill.

8. Enter a **25'** Right of Way.

9. Click <<**OK>>** to save your changes.

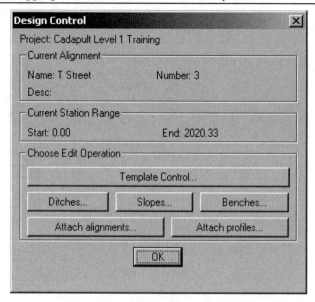

10. Click **<<OK>>** from the Design Control dialog box to process the sections.

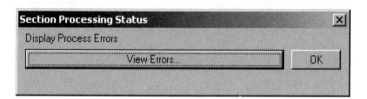

11. Click the **<<View Errors>>** button to view any errors.

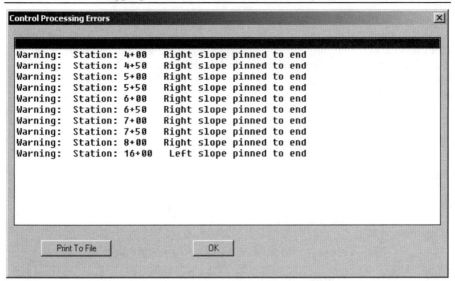

The **Pinned to end** error means that the cross section was not wide enough to daylight with the existing ground at the slope we defined in the design control. So Land Desktop has pinned the catch point for these sections at the edge of the section. To fix this and get the desired slope we will either need to sample wider sections or edit the finished ground centerline profile so that we are not cutting or filing as much in these areas. For our exercise we will accept the results with the catch point pinned to the end of the section.

12. Click **<<OK>>** to end the command.

13. View the results through **Cross Sections** ⇒ **View/Edit Sections.**

The road design is now completed. You can extract information from it to complete you construction documents like setting COGO points and volumes for earthworks as well as the materials that were defined in the material table.

8.5   Template Volumes

Volumes can be calculated from our finished road design. First we will calculate the total volume or earthwork volumes. Then we will calculate the template surface volume, which are the volumes from the materials in our material table.

8.5.1   Total Earthwork Volumes

1. Create a new layer called "Volume Table" and set it current.

2. Select **Cross Sections** ⇒ **Total Volume Output** ⇒ **Volume Table**.

3. Type **A** to do an Average End Area calculation.

4. Use **Curve Correction.**

5. No Volume Adjustment Factors.

6. Accept the default beginning and ending stations.

7. Pick the insertion point in a blank area of the drawing

Now you have calculated the grading volumes. Next we will calculate the volume of the materials as defined by the surfaces in our template.

8.5.2   Template Surface Volumes

1. Select **Cross Sections** ⇒ **Surface Volume Output** ⇒ **Template Surface.**

2. Type **A** to do an Average End Area calculation.

3. Use **Curve Correction.**

4. No Volume Adjustment Factors.

5. Use the file name **C:\T-street-vol.txt**

6. Accept the default beginning and ending stations.

Open **C:\T-street-vol.txt** in NotePad to review the results.

8.6   Setting Points From A Template

Points can be created from our Cross Section design for use as surface data or construction staking.

8.6.1   Adding Description Keys

To place the points on the desired layers you will need to define Description Keys. We will add the Description Keys to the "Training" file.

1. Select **Points** ⇒ **Point Management** ⇒ **Description Key Manager**.

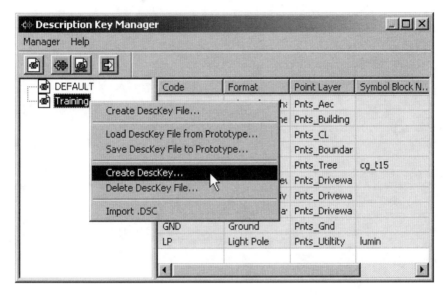

2. **Right-click** on the **Training** description key file and select ⇒ **Create DescKey**.

3. Add the Description Keys listed below:

Code	DESCRIPTION	LAYER
Center*	CENTER LINE	PNTS-CL
Catch	CATCH	PNTS-CATCH
EP	EDGE OF PAVEMENT	PNTS-EP
FL	FLOW LINE	PNTS-FL
BOC	BACK OF CURB	PNTS-BOC
TOC	TOP OF CURB	PNTS-TOC
SW	SIDEWALK	PNTS-SW

8.6.2   Point Settings

1.   Select **Points ⇒ Point Settings**.

2.   Select the **Create Tab.**

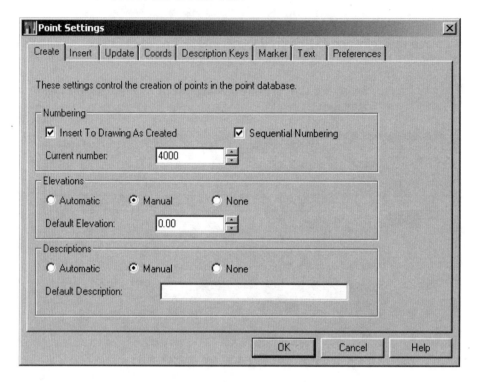

3.   Set the current Point Number to **4000.**

4. Select the **Insert Tab**.

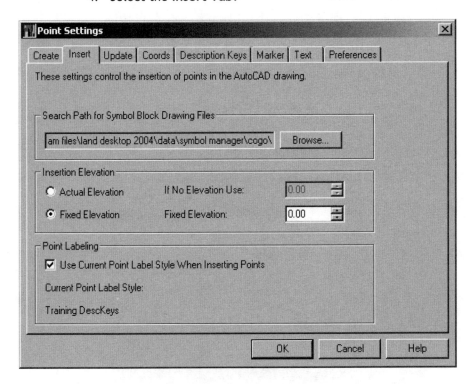

5. Confirm that **Point Labeling** is using the current Point Label Style and that the Current Point Label Style is set to **Training Desckeys**.

Unfortunately, if the Current Point Label Style is not set correctly you cannot change it here. You must go to the Labels pull-down and select Label Settings or change it on the Style Properties toolbar. This may require you to switch to the Land Desktop menu palette as well.

6. Click **<<OK>>** to save the changes to the Point Settings.

8.6.3   Setting Points From Point Codes

1. Select **Cross Sections** ⇒ **Point Output** ⇒ **Tplate Points to DWG**.

2. Accept the default beginning and ending stations.

3. Type **P** to use the Point Codes.

4. Confirm that the **Level 1 Training** point code table is set current if you are prompted.

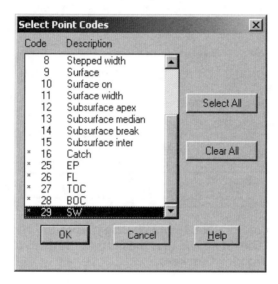

5. Select the Point Codes **Centerline**, **Catch**, **EP**, **FL**, **TOC**, **BOC** and **SW**.

The selection method here is not a typical windows selection method where you use the Ctrl or Shift keys while picking from the list to pick multiple items or ranges. In this dialog box picking an item once puts a * beside it showing that the item is selected and picking it twice removes the * deselecting the item. The fact that an item is highlighted means nothing.

6. Click **<<OK>>**.

7. Use **4000** as the current point number.

The points are now created in the project and inserted into the drawing. These points can be used in a point group as surface data to create a surface and generate contours or exported for construction staking.

8.7   A Grid Of 3D Faces

You can also import a grid of 3D Faces. This gives you a 3D model of your road that you can visually check in the Object Viewer. You can also build a surface from those 3D faces.

8.7.1   Importing A Grid Of 3D Faces

1. Select **Cross Sections** ⇒ **3D Grid**.

2. Accept the default beginning and ending stations.

3. Type **T** to make the grid from the Top Surface.

4. Use Top Surface **1**.

5. Vertical Scale Factor of **1**.

6. The Base Elevation is **0**.

The 3D Grid is placed on layer Rdgrid.

7. Isolate layer "**RDGRID.**"

8. Select all of the 3D faces with a crossing window.

9. **Right-click** and select ⇒ **Object Viewer.**

10. Rotate the grid in the Object Viewer to view our finished road from any angle you desire.

11. Close the Object Viewer.

A finished ground surface can be built from the Template Points that you imported or from the 3D faces that were imported from the top surface of the template. In the following exercise we will build the surface from the 3D faces. If you chose to build your surface from the template points instead make sure that you don't forget to use a Point Group.

8.7.2   Building a Surface from a 3D Grid

1.   **Isolate** the layer **RDGRID** if it is not already.

Next you will create a boundary for the surface by importing the daylight lines.

2.   Select **Cross Sections** ⇒ **Point Output** ⇒ **Catch Points to DWG**.

3.   When asked to import the catch points answer **NO**.

4.   When asked to import the Daylight Lines answer **YES**.

5.   Accept the defaults for beginning and ending stations.

This will give you line segments between the catch points which is also the boundary of the surface we are going to create.

6.   **Isolate** the layer **Daylight** and set it **Current**.

7.   Fill in the gaps at the ends of the alignment and at the intersection using the **Pline** command and an **endpoint snap**.

8.   Join the lines into one closed polyline that will become our surface boundary.

9.   Select **Terrain** ⇒ **Terrain Model Explorer**.

10. Right-click on the **Terrain folder** and select ⇒ **Create New Surface**.

11. Rename the surface to "**T Street**".

12. In the **Terrain Model Explorer** under the surface **T Street** right-click on **Boundaries** and select ⇒ **Add Boundary Definition**.

13. Pick the Polyline you created on the Daylight layer.

14. Accept the default for the Boundary Name. The boundary Type is **Outer**. Do not make breaklines along the boundary.

15. Turn **on** the layer **RDGRID**.

16. Under the surface **T Street** right-click on **Point Files** ⇒ **Add Points from AutoCAD Objects** ⇒ **3d Faces**.

17. Select objects by **Layer** and pick one of the 3D faces on the road surface. **ENTER** to end the selection.

Note: A list of the x,y,z values of each corner of all the 3D faces is created. This will be our surface data.

18. In the **Terrain Model Explorer** right-click on **T Street** and select ⇒ **Build**. Confirm the options for using Point File Data and Apply Boundaries are checked, and click **<<OK>>**.

Now that the surface is built you can Import the 3D Lines and Flip Faces and Delete Lines as needed using the Commands in **Terrain** ⇒ **Edit Surface**.

8.8   Road Output

The Road Output Commands allow you to create surfaces, draw 3D polylines from point codes, and draw 3D daylight lines directly from your template design. This is a new alternative to some of the commands that we looked at in previous exercises and may be the best option for many projects. However, it is important to understand the process of some of the longer methods such as building a surface from points or 3D faces and importing daylight lines because you may need to use them for portions of more complicated designs.

8.8.1 Creating A Road Surface

The Create Road Surface command automates the process of building a surface from the top surface of a template. It basically condenses the two previous exercises into one dialog box.

1. Select **Cross Sections** ⇒ **Road Output** ⇒ **Create Road Surface**.

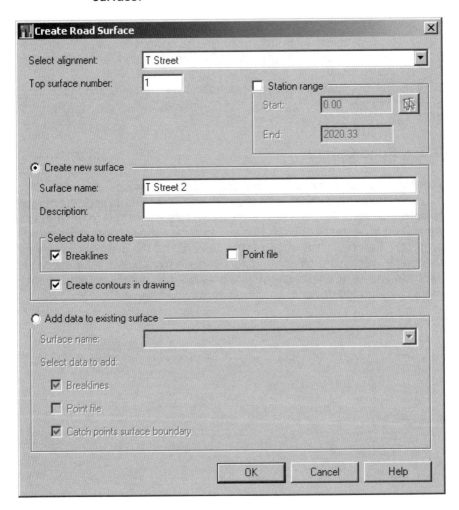

2. Select the alignment **T Street**.

3. Set the **Top Surface** number to **1**.

4. Name the new surface **T Street 2**.

5. Create **Breakline** data from the cross sections to be used as our surface data.

6. **Enable** the option to automatically **Create Contours** from the new surface.

7. Click **<<OK>>** to build the surface.

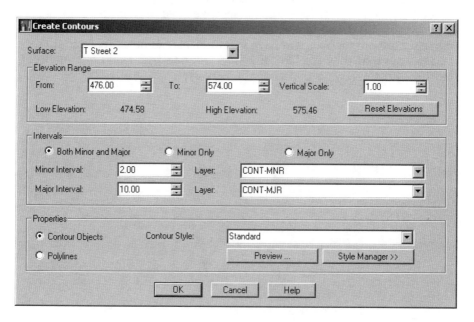

8. Click **<<OK>>** to create the contours from the new surface.

9. Answer **Yes** to **Erase** the old contours when prompted.

Now that the surface has been built that represents your road design, whether it has been built from a Point Group, 3D faces, or the Road Output command, you can paste it into a copy of the existing ground surface to create a surface that represents the final site conditions of the design. Then you perform any of the surface display and analysis commands that we used in previous chapters.

8.9   Chapter Summary

In this chapter we created a Template, applied it through Design Control, and extracted design data for volumes, points, and a surface of the final site conditions.

Now that you have worked with the Profile and Cross Section commands take a few minutes to think about other types of projects in which you might be able to use these commands to make yourself more efficient. Remember, all of the different commands we have used in this book are just tools for you to use in your projects any way that you can to make yourself more productive. We have shown many examples that use these tools in ways that they are typically used, but don't let that limit you from thinking of them in ways that could be very useful to you.